"This is truly an amazing book! The product of a unique collaboration between a renowned economist and renowned seismologist (who happen to be father and son), *Playing against Nature* lays out a clear story, in easy-to-read prose, of what natural disasters are, what the limitations of risk prediction can be, and how society's response to them has to account for the reality that we have limited economic resources. The authors present fascinating case studies to illustrate examples of where predictions have failed, and why. They also take a bold step by showing how natural disasters and economic disasters provide similar challenges, and provide a clear description of how risk should be assessed, and how it can be mitigated reasonably. This is a book that researchers, policy makers, and the general public should read. It can even serve as valuable text for the new generation of interdisciplinary college courses addressing the interface between science and social science." – Stephen Marshak, Professor and Director of the School of Earth Society and Environment, University of Illinois at Urbana-Champaign

"I very highly recommend this book for anyone dealing with or interested in natural hazards assessment and mitigation. It is a *tour de force* with examples, descriptions, illustrations, reference lists, and explanations for understanding natural disasters and negotiating the often perilous and misguided approaches for hazards mitigation. This book is a huge achievement in that it has collected an enormous amount of relevant information, case studies, economics and engineering factors, loss statistics, references, and even study guides and questions for students. It is both highly technical with all the probability and statistics formulations needed to express necessary relationships but on the other hand, so well written that professionals in government, business, and education will find it exceedingly readable. In my everyday work experience, I attempt to communicate principles of hazard occurrences and risks. This book gives me far more useable material than I have ever had to achieve my goals for advising public officials, teaching university students, and educating citizens. This is the best resource in existence for understanding natural hazards and hazard mitigation." – James C. Cobb, State Geologist and Director, Kentucky Geological Survey, University of Kentucky

"*Playing against Nature* is a virtuoso performance by a father-son duo. A distinguished economist and seismologist have produced a pioneering work that promises to enhance our ability to integrate assessment science, cost-benefit analysis and mitigation design and engineering. The result will be more informed, bottom-up, hazard mitigation policies. This outstandingly researched book is highly readable and destined to become a classic." – Steve H. Hanke, Professor of Applied Economics, The Johns Hopkins University

"Elegantly written in Seth Stein's usual memorable prose, *Playing against Nature* treats jointly seismic and economic catastrophes in a thought-provoking and readable way. How blindingly obvious something can be after the event! Ringing oh so very true, it provides insight into why science and scientists don't get things right all the time. Enriched with gems of quotes, and an unusual mix of hard science and philosophy, *Playing against Nature* will make a great supporting text for any course on hazards – geologic, engineering, political or economic – and judging from current trends, we could all use as much understanding of this topic as possible." – Gillian R. Foulger, Professor of Geophysics, University of Durham

"Authored by a remarkable father and son team, *Playing against Nature* is a comprehensive, lucid assessment of the interplay between natural hazards and economics of many kinds. As world population continues to increase to more and more unsustainable numbers, and demand for economic growth plagues the world, human activities continue to place us in more and greater vulnerability as Earth processes go on, as they have over deep time. We need to better recognize and thus more responsibly prepare for inevitable natural events. Blunt, forceful, and true statements (e.g., 'Humans have to live with natural hazards' and 'Hazards are geological facts that are not under human control') characterize *Playing against Nature* and make reading this contribution, by anyone, a sobering and enlightening experience. I highly recommend *Playing against Nature* to those who care about the future of the human race." – John Geissman, Professor of Geosciences, University of Texas at Dallas

"In the wake of recent natural disasters and economic crises, the authors question the inability of specialists – of earth and planetary sciences on one side and economists on the other – to predict such events. Beyond these two spheres, this work also reveals a bridge between seemingly distinct fields of science, which meet as soon as one starts to focus on concepts that are fundamental for both, such as hazard, risk or vulnerability. This book discusses the laws of probability and the most appropriate models for predicting rare events; it also offers strategies to optimize mitigation plans. *Playing against Nature* thus is an innovative work that should encourage researchers in different disciplines to collaborate. It may also become a useful tool for graduate students. This book furthermore constitutes an ideal reference work for policy makers." – Serge Rey, Professor of Economics, University of Pau

"Insightful and provocative, *Playing against Nature* by Stein and Stein explains in a brilliant yet playful way why experts missed many of the recent natural and manmade disasters, from the 2011 Tohoku earthquake to the 2008 financial crisis. It makes an enjoyable read for anyone who has ever wondered how society prepares and responds to natural disasters. The authors, an economist father and a geophysicist son, provide a unique perspective of how scientific study of natural disasters interplays with policy making for hazard mitigation. As a student of earthquake science, I found many arguments and facts in the book compelling and intriguing. Facing many unknowns and with limited resources, we are gambling with nature in hazard preparation and mitigation, as the authors put it. We may not expect to win every hand, but we need to understand our odds. *Playing against Nature* offers a fresh way to look at nature's games. It should be helpful to professionals, and delightful to everyone who opens the book." – Mian Liu, Curators' Distinguished Professor in Geological Sciences, University of Missouri

 "How can policy defend society better against natural disasters whose probabilities are uncertain and in flux? In *Playing against Nature*, Seth Stein, a geologist, and his late father Jerome, an economist, joined forces. Their book is a clear Guide for the Perplexed, combining scholarship and exposition to show how to prepare more wisely for hurricanes, earthquakes, and tsunamis." – Shlomo Maital, Professor Emeritus, Institute for Advanced Studies in Science and Technology, Technion-Israel Institute of Technology

"What do natural disasters and economic disasters have in common, and how is it possible to efficiently mitigate their effects? You will find the answer in this scholarly book. But there is more to it than meets the eye: this important monograph is based on what I call 'the Steins' synergy' (after the late Jerome Stein, an economist, and his son Seth Stein, a geoscientist). The interaction between these two scientists has been such that the combined result of their joint research, reported in this book, is much greater than the sum of the individual results: the quintessential example of what interdisciplinarity can achieve." – Giancarlo Gandolfo, Professor, Accademia Nazionale dei Lincei, Rome, Research Fellow, CESifo, Munich, Professor of International Economics, Sapienza University of Rome (retired)

" 'Nature's smarter than us' might be a good subtitle for this well-written and illustrated tome by a father-son team. Reviewing numerous natural disasters from Katrina to Haiti to Sandy to the Japan earthquake, the authors find most disaster responses to be seriously wanting. Their accounts of nature at its most violent range from humorous to appalling. The solution: a better understanding of the uncertainties of disaster response, free of politics, tradition and too narrow science." – Orrin H. Pilkey, Professor Emeritus of Earth and Ocean Sciences, Duke University

Playing against Nature

Jerome and Seth Stein, spring 2012. Photo by Hadassah Stein.

Playing against Nature

Integrating Science and Economics to Mitigate Natural Hazards in an Uncertain World

Seth Stein[1] *and Jerome Stein*[2]

[1]Department of Earth and Planetary Sciences, Northwestern University, Evanston, Illinois
[2]Division of Applied Mathematics, Brown University, Providence, Rhode Island

This work is a co-publication between the American Geophysical Union and Wiley

 WILEY

This edition first published 2014 © 2014 by John Wiley & Sons, Ltd
This work is a co-publication between the American Geophysical Union and Wiley

Registered office: John Wiley & Sons, Ltd, The Atrium, Southern Gate, Chichester, West Sussex, PO19 8SQ, UK

Editorial offices: 9600 Garsington Road, Oxford, OX4 2DQ, UK
The Atrium, Southern Gate, Chichester, West Sussex, PO19 8SQ, UK
111 River Street, Hoboken, NJ 07030-5774, USA

For details of our global editorial offices, for customer services and for information about how to apply for permission to reuse the copyright material in this book please see our website at www.wiley.com/wiley-blackwell.

Library of Congress Cataloging-in-Publication Data

Stein, Seth.
  Playing against nature : integrating science and economics to mitigate natural hazards in an uncertain world / Seth Stein and Jerome Stein.
     pages cm
  Includes bibliographical references and index.
  ISBN 978-1-118-62082-3 (cloth)
1. Hazard mitigation. 2. Hazard mitigation–Economic aspects. 3. Natural disasters–Economic aspects. 4. Natural disasters–Risk assessment. 5. Emergency management. I. Stein, Jerome L. II. Title.
  GB5014.S83 2014
  363.34'6–dc23
                                  2013046062

A catalogue record for this book is available from the British Library.

Wiley also publishes its books in a variety of electronic formats. Some content that appears in print may not be available in electronic books.

Cover image: Photo 1 from Moment of the tsunami slideshow. A wave approaches Miyako City from the Heigawa estuary in Iwate Prefecture after the magnitude 8.9 earthquake struck the area, March 11, 2011. Reuters/Mainichi Shimbun.
Cover design by Steve Thompson

Set in 10.5/12.5 pt TimesLTStd by Toppan Best-set Premedia Limited

Printed and bound in Singapore by Markono Print Media Pte Ltd

1   2014

Contents

Preface

This book considers how to make policy to defend society against natural hazards more effective. Recent events including Hurricane Katrina in 2005, the 2011 Tohoku earthquake and tsunami, and Hurricane Sandy in 2012 show that in its high-stakes game of chance against nature, society often does poorly. Sometimes nature surprises us, when an earthquake, hurricane, or flood is bigger or has greater effects than expected from detailed natural hazard assessments. In other cases, nature outsmarts us, doing great damage despite expensive mitigation measures being in place, or causing us to divert limited resources to mitigate hazards that are overestimated.

This situation may seem surprising because of the steady advances being made in the science of natural hazards. In our view, much of the problem comes from the fact that formulating effective natural hazard policy involves using a complicated combination of science and economics to analyze a problem and explore the costs and benefits of different options, in situations where the future is very uncertain. In general, mitigation policies are chosen without this kind of analysis. Typically, communities have not looked at different options, and somehow end up choosing one or having one chosen for them without knowing how much they're paying or what they're getting for their money. This is like buying insurance without considering how much a policy will cost and what the benefits would be. Not surprisingly, the results are often disappointing. Thus it is worth thinking about how to do better.

This book explores these issues, taking a joint view from geoscience and economics. My view is that of a seismologist interested in the science of large earthquakes and earthquake hazard mitigation. My coauthor and late father, Jerome Stein, brought the view of an economist interested in public policy.

As my father told the *Brown Daily Herald* in November, 2012, he viewed this book as derived from the day in 1960 that he took his 7-year old son to hear a lecture about the new discoveries of continental drift that would soon transform modern geology. Apparently I was intrigued by the idea, and asked the speaker whether drifting continents were like bars of soap floating in the bathtub.

Over the years, my father and I often talked about science. We discussed natural hazards, starting in 1998, when I became skeptical of widely-touted claims that parts of the central US faced earthquake hazards as high as California's, and that buildings should be built to the same safety standards. To my surprise, it turned out that the federal government was pressing for these measures without undertaking any analysis of the huge uncertainties in the hazard estimates or of whether the large costs involved would yield commensurate benefits to public safety. To my further surprise, my father said that this was typical, in that economists had found that many health and safety regulations were developed without such analysis or consideration of alternative policies. In such cases, no one knew whether these policies made sense or not. I became interested in this question, and started working with colleagues and students to investigate how large the uncertainties in earthquake hazard estimates were.

Our discussions on this topic ramped up in 2011, following the Tohoku earthquake and tsunami. Japanese hazard planners had assumed that an earthquake and tsunami that big could not occur there, whereas my colleague Emile Okal and I had found before the earthquake that they could. At the same time, my father was studying how the 2008 US financial disaster had occurred, despite the fact that both Wall Street and the US government had been sure – based on economic models – that it could not. We realized that although one disaster was natural and the other was economic, they had much in common. Both resulted from overconfidence in how well hazards could be assessed, both had vulnerabilities that were not recognized, and the result in both cases was poorly formulated policies.

We decided to explore these issues in a series of journal articles that became the basis of this book. Because there are many fine books on natural hazard science and on economics, we focused on the interface between the two fields. Our discussions of the challenging questions involved and how to present them had special intensity because we started the book after my father's illness was diagnosed and knew we had only a short time to finish it.

For simplicity, we decided to primarily use earthquake and tsunami hazards as examples, although the approach applies to other natural hazards. Our goal is to introduce some key concepts and challenges, and illustrate them with examples and questions that we pose at the end of each chapter. We decided to introduce some relevant mathematics, which can be skipped by readers without losing the key themes. We illustrate the key themes with examples and questions at the end of each chapter. As is typical for natural hazards, many of the questions are difficult and few have unique or correct answers.

In this sprit of looking toward the future, we hope the book will help researchers, especially younger ones, to develop an interdisciplinary outlook as they work at the interface between the two fields. Hopefully their work, both about hazards and how to make better policies, will help society fare better in its game against nature.

Seth Stein
Glencoe, Illinois
April 2013

Royalties from this book go to the Division of Applied Mathematics at Brown University to support the Jerome L Stein award, which recognizes undergraduate students who show outstanding potential in an interdisciplinary area that involves applied mathematics.

Acknowledgments

Although science is always a human endeavor, this book is especially so because of its father–son collaboration. It would not have been completed, given my father's illness, without the support of Hadassah and Carol Stein. Their encouragement when the task seemed too big and progress slowed is even more impressive given that both went through it all for our previous books.

This book grew from ideas developed over many years via research carried out with coworkers, fruitful discussions with them and other researchers studying these or related problems, and knowledge from the broad communities of geoscientists, economists, and others interested in natural hazards. In that spirit, I would like to thank many people. All should feel free to share credit for parts of the book with which they agree, and disclaim parts with which they disagree.

I thank my coauthors on the research papers discussed here, including Eric Calais, Carl Ebeling, Robert Geller, Richard Gordon, James Hebden, Qingsong Li, Mian Liu, Jason McKenna, Andres Mendez, Andrew Newman, Emile Okal, Carol Stein, John Schneider, Laura Swafford, and Joe Tomasello.

I have also have benefited from discussions with John Adams, Amotz Agnon, Rick Aster, Roger Bilham, Eric Calais, Thiery Camelbeeck, Dan Clark, Nick Clark, Sierd Cloetingh, Jim Cobb, Mike Craymer, Bill Dickinson, Tim Dixon, Roy Dokka, Joe Engeln, Andy Freed, Tom James, John Geissman, Bob Hermann, David Hindle, Tom Holzer, Sue Hough, Ken Hudnut, Alex Forte, Anke Friedrich, Alan Kafka, Steve Kirby, Jonas Kley, Cinna Lomnitz, Mike Lynch, Steve Marshak, Ailin Mao, Glenn Mattiolli, Miguel Merino, Brian Mitchell, Antonella Peresan, Orrin Pilkey, Hans-Peter Plag, Paul Rydelek, Jim Savage, Gary Searer, Giovanni Sella, Norm Sleep, Bob Smalley, Bob Smith, Bruce Spenser, Ross Stein, Roy van Arsdale, David Wald,

Zhenming Wang, John Weber, Steve Wesnousky, Rinus Wortel, Michael Wysession, Dave Yuen, and many others.

I thank Gillian Foulger, Peter Clark, Serge Rey, Hadassah Stein and an anonymous reader for reviewing the manuscript. Finally, I thank the Alexander von Humboldt Foundation for supporting my stay at the University of Gottingen, where the book's editing was completed.

Note on Further Reading and Sources

Natural hazards and disasters are so dramatic that a wealth of information is easily available. One source is introductory texts. Another is general audience books about specific disasters such as the 1906 San Francisco earthquake, the 2004 Indian Ocean tsunami, or Hurricane Katrina. The World-Wide Web has lots of information about individual disasters, including news stories, photographs, and video. Information on the Web is convenient but variable in quality. That on technical topics, such as high-precision GPS or earthquake-resistant construction, is often excellent. In addition, many primary sources such as the Japanese parliament's Fukushima nuclear accident commission report or the American Society of Civil Engineers report about Hurricane Katrina are available online. However, because websites are easily created and copied from each other, some contain information that is wrong or out of date. For example, a Google search found more than 32,000 references, including the online encyclopedia Wikipedia, to the incorrect legend that the 1811–1812 New Madrid earthquakes rang church bells in Boston.

Technical information on the scientific topics discussed here is often more easily accessible from textbooks than from research papers written tersely by scientists for scientists familiar with the topics under discussion. We list several textbooks for specific chapters. Research papers mentioned, including those from which a figure is used, are listed in the references by their authors.

The scientist has a lot of experience with ignorance and doubt and uncertainty, and this experience is of very great importance, I think. When a scientist does not know the answer to a problem, he is ignorant. When he has a hunch as to what the result is, he is uncertain. And when he is pretty damn sure of what the result is going to be, he is still in some doubt. We have found it of paramount importance that in order to progress, we must recognize our ignorance and leave room for doubt.

Richard Feynman, 1988

About the Companion Website

This book is accompanied by a companion website:

www.wiley.com/go/stein/nature

The website includes:

- Powerpoints of all figures from the book for downloading
- PDFs of tables from the book

1

A Tricky, High-Stakes Game

Earthquake risk is a game of chance of which we do not know all the rules. It is true that we gamble against our will, but this doesn't make it less of a game.

Lomnitz (1989)[1]

1.1   Where We Are Today

Natural hazards are the price we pay for living on an active planet. The tectonic plate subduction producing Japan's rugged Tohoku coast gives rise to earthquakes and tsunamis. Florida's warm sunny weather results from the processes in the ocean and atmosphere that cause hurricanes. The volcanoes that produced Hawaii's spectacular islands sometimes threaten people. Rivers that provide the water for the farms that feed us sometimes flood.

Humans have to live with natural hazards. We describe this challenge in terms of *hazards*, the natural occurrence of earthquakes or other phenomena, and the *risks*, or dangers they pose to lives and property. In this formulation, the risk is the product of hazard and *vulnerability*. We want to *assess* the hazards – estimate how significant they are – and develop methods to *mitigate* or reduce the resulting losses.

Hazards are geological facts that are not under human control. All we can do is try to assess them as best we can. In contrast, risks are affected by human actions that increase or decrease vulnerability, such as where people live and

[1] Lomnitz, 1989. Reproduced with permission of the Seismological Society of America.

Playing against Nature: Integrating Science and Economics to Mitigate Natural Hazards in an Uncertain World, First Edition. Seth Stein and Jerome Stein.
© 2014 John Wiley & Sons, Ltd. Published 2014 by John Wiley & Sons, Ltd.
Companion Website: www.wiley.com/go/stein/nature

how they build. We increase vulnerability by building in hazardous areas, and decrease it by making buildings more hazard resistant. Areas with high hazard can have low risk because few people live there. Areas of modest hazard can have high risk due to large population and poor construction. A disaster occurs when – owing to high vulnerability – a natural event has major consequences for society.

The harm from natural disasters is enormous. On average, about 100,000 people per year are killed by natural disasters, with some disasters – such as the 2004 Indian Ocean tsunami – causing many more deaths. Although the actual numbers of deaths in many events, such as the 2010 Haiti earthquake, are poorly known, they are very large.

Economic impacts are even harder to quantify, and various measures are used to try to do so. Disasters cause *losses*, which are the total negative economic impact. These include direct losses due to destruction of physical assets such as buildings, farmland, forests, etc., and indirect losses that result from the direct losses. Because losses are hard to determine, what is reported is often the *cost*, which refers to payouts by insurers (called *insured losses*) or governments to reimburse some of the losses. Thus the reported cost does not reflect the losses to people who do not receive such payments. Losses due to natural disasters in 2012 worldwide are estimated as exceeding $170 billion (Figure 1.1). Damages within the US alone cost insurers about $58 billion. Disaster losses are on an increasing trend, because more people live in hazardous areas. For example, the population of hurricane-prone Florida has grown from 3 million in 1950 to 19 million today.

Society can thus be viewed as playing a high-stakes game of chance against nature. We know that we will lose, in two ways. If disaster strikes, direct and indirect losses result. In addition, the resources used for measures that we hope will mitigate the hazards and thus reduce losses in the future are also lost to society, because they cannot be used for other purposes.

Thus the challenge is deciding *how much mitigation is enough*. More mitigation can reduce losses in possible future disasters, at increased cost. To take it to the extreme, too much mitigation could cost more than the problem we want to mitigate. On the other side, less mitigation reduces costs, but can increase potential losses. Hence too little mitigation can cause losses that it would make more sense to avoid. We want to hit a "sweet spot" – a sensible balance. This means being careful, thoughtful gamblers.

We want to help society to come up with strategies to minimize the combined losses from disasters themselves and from efforts to mitigate them. This involves developing methods to better assess future hazards and mitigate their effects. Because both of these are difficult, our record is mixed. Sometimes we do well, and sometimes not.

(a)

Natural Catastrophes 2012
World map

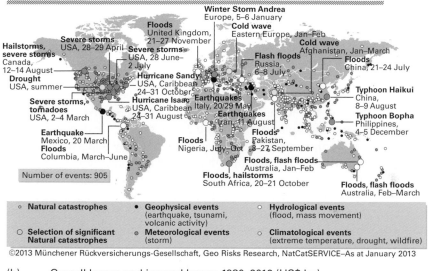

Number of events: 905

○ **Natural catastrophes**	● **Geophysical events** (earthquake, tsunami, volcanic activity)	○ **Hydrological events** (flood, mass movement)
○ **Selection of significant** **Natural catastrophes**	◉ **Meteorological events** (storm)	○ **Climatological events** (extreme temperature, drought, wildfire)

©2013 Münchener Rückversicherungs-Gesellschaft, Geo Risks Research, NatCatSERVICE–As at January 2013

(b)   Overall losses and insured losses 1980–2012 (US$ bn)

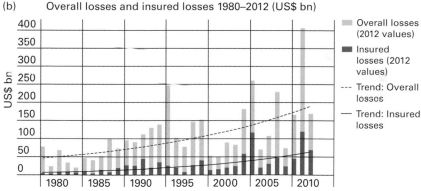

Figure 1.1 (a) Natural disasters in 2012. (Munich Re, 2013a. Reproduced with permission from Munich Reinsurance Company AG.) (b) Overall and insured losses since 1980 due to natural disasters. (Munich Re, 2013b. Reproduced with permission from Munich Reinsurance Company AG.)

On the hazard assessment side, the problem is that we lack full information. Geoscience tells us a lot about the natural processes that cause hazards, but not everything. We are learning more by using new ideas and methods that generate new data, but still we have a long way to go. For example, meteorologists are steadily improving forecasts of the tracks of hurricanes, but forecasting their strength is harder. We know a reasonable amount about why and where earthquakes will happen, have some idea about how big they will be, but much less about when they will happen. We thus need to decide what to do given these uncertainties.

This situation is like playing the card game of blackjack, also called "21." Unlike most other card games, blackjack is considered more a game of skill than a game of chance. As mathematician Edward Thorp showed, despite the randomness in the cards drawn, skilled players can on average win by a small fraction using a strategy based on the history of the cards that have already been played. MIT student blackjack teams using these winning strategies formed the basis of the fictionalized 2008 film "21." A key aspect of the game is that players see only some of the casino dealer's cards. Dealing with natural hazards has the further complication that we do not fully understand the rules of the game, and are trying to figure them out while playing it.

On the mitigation side, methods are getting better and cheaper. Still, choosing strategies is constrained because society has finite resources. There's no free lunch – resources used for mitigating hazards are not available for other purposes. Funds spent by hospitals to strengthen buildings to resist earthquake shaking cannot be used to treat patients. Money spent putting more steel in school buildings does not get used to hire teachers. Spending on seawalls and levees comes at the expense of other needs. Choosing priorities is always hard, but it is especially difficult when dealing with natural hazards, because of our limited ability to forecast the future.

When natural hazard planning works well, hazards are successfully assessed and mitigated, and damage is minor. Conversely, if a hazard is inadequately mitigated, sometimes because it was not assessed adequately, disasters happen. Disasters thus regularly remind us of how hard it is to assess natural hazards and make effective mitigation policies. The earth is complicated, and often surprises or outsmarts us. Thus although hindsight is always easier than foresight, examining what went wrong points out what we should try to do better.

The effects of Hurricane Katrina, which struck the US Gulf coast in August 2005, had been anticipated. Since 1722, the region had been stuck by 45 hurricanes. As a result, the hazard due to both high winds and flooding of low-lying areas including much of New Orleans was recognized. Mitigation measures including levees and flood walls were in place, but recognized to be inadequate to withstand a major hurricane. It was also recognized that many New Orleans residents who did not have cars would likely not be able to evacuate unless procedures were established. Thus despite accurate and timely warning by the National Weather Service as the storm approached, about 1,800 people died. The total cost of the damage caused by the disaster is estimated at $108 billion, making Katrina the costliest hurricane in US history.

Japan has a major earthquake problem, illustrated by the 1923 Kanto earthquake that caused more than 100,000 deaths in the Tokyo region. Hence

Figure 1.2 More than a dozen ships were washed inland by the Tohoku tsunami in Kesennuma City, Miyagi Prefecture. The fishing trawler Kyotoku-maru came to rest on a giant debris pile on one of the main roads to City Hall. (Courtesy of Hermann M. Fritz.)

scientists have studied the Japanese subduction zone extensively for many years using sophisticated equipment and methods, and engineers have used the results to develop expensive mitigation measures. But the great earthquake that struck Japan's Tohoku coast on March 11, 2011 was much larger than predicted even by sophisticated hazard models, and so caused a tsunami that overtopped giant seawalls (Figure 1.2). Although some of the mitigation measures significantly reduced losses of life and property, the earthquake caused more than 15,000 deaths and damage costs of $210 billion.

After the Tohoku earthquake the immediate question that arose was if and how coastal defenses should be rebuilt: the defences had fared poorly and building mitigation measures to withstand tsunamis as large as the one on March 2011 is too expensive. A similar issue soon arose along the Nankai Trough to the south, where new estimates warning of giant tsunamis 2–5 times higher than in previous models (Figure 1.3) raised the question of what to do, given that the timescale on which such events may occur is unknown and likely to be of order 1000 years. In one commentator's words, "the question is whether the bureaucratic instinct to avoid any risk of future criticism by presenting the worst case scenario is really helpful . . . What can (or should be) done? Thirty meter seawalls do not seem to be the answer."

The policy question, in the words of Japanese economist H. Hori, is:

What should we do in face of uncertainty? Some say we should spend our resources on present problems instead of wasting them on things whose results

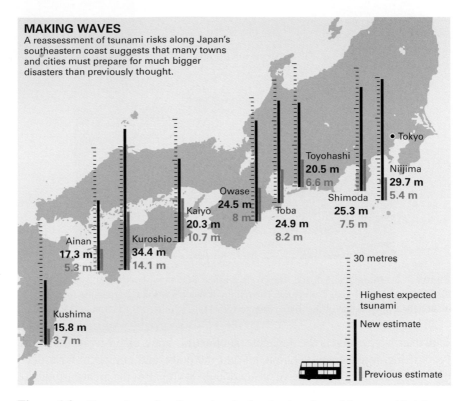

Figure 1.3   Comparison of earlier and revised estimates of possible tsunami heights from a giant Nankai Trough earthquake (Cyranoski, 2012a. Reproduced with permission from *Nature*.)

are uncertain. Others say we should prepare for future unknown disasters precisely because they are uncertain.

1.2   What We Need to Do Better

The Tohoku earthquake was the "perfect storm," illustrating the limits of both hazard assessment and mitigation, and bringing out two challenges that are the heart of this book. We discuss them using earthquakes as examples, but they arise for all natural hazards.

The first challenge is improving our ability to assess future hazards. It was already becoming clear that the methods currently used for earthquakes often fail. Tohoku was not unusual in this regard – highly destructive earthquakes, like the one in Wenchuan, China, in 2008, often occur in areas predicted by hazard maps to be relatively safe.

Another example is the devastating magnitude 7.1 earthquake that struck Haiti in 2010. As shown in Figure 1.4, the earthquake occurred where a hazard map made in 2001 predicted that the maximum ground shaking expected to

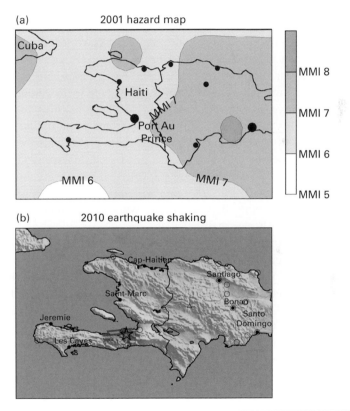

PERCEIVED SHAKING	Not felt	Weak	Light	Moderate	Strong	Very strong	Severe	Violent	Extreme
POTENTIAL DAMAGE	none	none	none	Very light	Light	Moderate	Moderate/Heavy	Heavy	Very Heavy
PEAK ACC.(%g)	<0.17	0.17–1.4	1.4–3.9	3.9–9.2	9.2–18	18–34	34–65	65–124	>124
PEAK VEL.(cm/s)	<0.1	0.1–1.1	1.1–3.4	3.4–8.1	8.1–16	16–31	31–60	60–116	>116
INSTRUMENTAL INTENSITY	I	II-III	IV	V	VI	VII	VIII	IX	X+

Figure 1.4   (a) Seismic hazard map for Haiti produced prior to the 2010 earthquake showing maximum shaking expected to have a 10% chance of being exceeded once in 50 years, or on average once about every 500 years. (b) Map of the shaking in the 2010 earthquake. (Stein et al., 2012. Reproduced with permission of Elsevier B.V.) See also color plate 1.4.

have a 10% chance of being exceeded once in 50 years, or on average once about every 500 ($= 50/0.1$) years, was intensity VI. Intensity is a descriptive scale of shaking, usually described by roman numerals, which we will discuss in Chapter 11. Intensity VI corresponds to strong shaking and light damage. Shaking is more precisely described by the acceleration of the ground, often as a fraction of "g," the acceleration of gravity ($9.8\,m/s^2$). Within ten years, much stronger shaking than expected – intensity IX, with violent shaking and heavy damage – occurred. Great loss of life also resulted, although estimates of the actual numbers of deaths vary widely.

The fundamental problem is that there is much we still do not know about where and when earthquakes are going to happen. A great deal of effort is being put into learning more – a major research task – but major advances will probably come slowly, given how complicated the earthquake process is and how much we do not yet understand. We keep learning the hard way to maintain humility before the complexity of nature. In particular, we are regularly reminded that where and when large earthquakes happen is more variable than we expected. Given the short geological history we have, it is not clear how to tell how often the biggest, rarest, and potentially most destructive earthquakes like the 2011 Tohoku one will happen. There are things we may never figure out, notably how to predict when big earthquakes will happen on any time scale shorter than decades.

Given this situation and the limitations of what we know, how can we assess hazards better today? The traditional approach to this problem is to make new hazard maps after large earthquakes occur in places where the map previously showed little hazard (Figure 1.5). This is an example of what statisticians call "Texas sharpshooting," because it is like first shooting at the barn and then drawing a target around the bullet holes.

To make things worse, sometimes the new map does not predict future earthquake shaking well and soon requires further updating. In Italy, for example, the national earthquake hazard map, which is supposed to forecast hazards over the next 500 years, has required remaking every few years (Figure 1.6).

Earthquake hazard mapping has become an accepted and widely used tool to help make major decisions. The problem is that although it seemed like a sensible approach, governments started using it enthusiastically before any careful assessment of the uncertainties in these maps or objective testing of how well they predict future earthquake shaking had been undertaken. Now that major problems are surfacing, we need to do better. One important task is to assess the uncertainties in hazard map predictions and communicate them to potential users, so that they can decide how much credence to place in the maps, and thus make them more useful. We also need to develop methods to

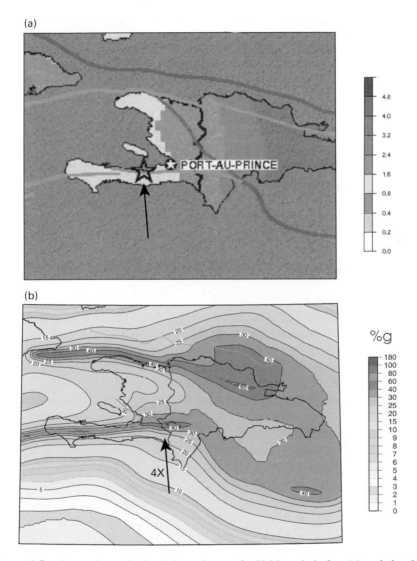

Figure 1.5   Comparison of seismic hazard maps for Haiti made before (a) and shortly after (b) the 2010 earthquake. The newer map shows a factor of four higher hazard on the fault that had recently broken in the earthquake. (Stein et al., 2012. Reproduced with permission of Elsevier B.V.) See also color plate 1.5.

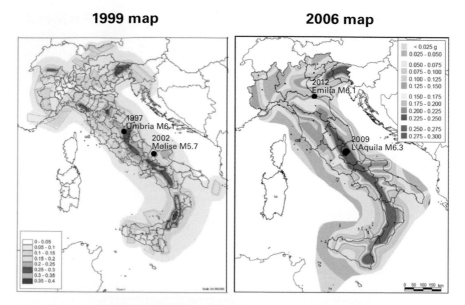

Figure 1.6   Comparison of successive Italian hazard maps, which forecast some earthquake locations well and others poorly. The 1999 map was updated after the missed 2002 Molise quake and the 2006 map will presumably be updated because it missed the 2012 Emilia earthquake. (Stein et al., 2013. Reproduced with permission of Elsevier B.V.) See also color plate 1.6.

objectively test these maps, to assess how well maps made with different methods describe what actually happens, and to improve future maps.

The second challenge is learning how to use what we know about hazards to develop mitigation policies. We need to develop sensible approaches to evaluate alternative strategies. In addition to science, this process involves complicated economic, societal, and political factors.

Typically, more extensive mitigation measures cost more, but are expected to further reduce losses in future events. For example, after Hurricane Katrina breached coastal defenses in 2005 and flooded much of New Orleans, choosing to what level these defenses should be rebuilt became an issue. Should they be rebuilt to withstand only a similar hurricane, or stronger ones? Similarly, given the damage to New York City by the storm surge from Hurricane Sandy in 2012, options under consideration range from doing little, through intermediate strategies such as providing doors to keep water out of vulnerable tunnels, to building up coastlines or installing barriers to keep the storm surge out of rivers.

Although our first instinct might be to protect ourselves as well as possible, reality sets in quickly, because resources used for hazard mitigation are not available for other societal needs. For example, does it make sense to spend billions of dollars making buildings in the central US as earthquake-resistant as in California, or would these funds do more good if used otherwise? Should all hospitals in California be made earthquake-resistant, or would it be wiser to use these resources caring for millions of people without health insurance? As a doctor mused, "we could treat a lot of people for $50 billion." In the same spirit, a European Union official charged with hazard mitigation pointed out that plans for higher levees to reduce river flood damage compete for funds with plans to improve kindergartens.

These difficult issues are discussed in an editorial "Quake work needs limits and balance" in the *New Zealand Herald* after the 2011 Christchurch earthquake that caused 158 deaths and considerable damage. In the newspaper's view,

> Mandatory quake-proofing of all New Zealand buildings would, however, be hugely expensive. Proponents say this would be worthwhile if even one life is saved, let alone the hundreds lost in Christchurch. But the need for preparedness must be balanced so as not to be out of all proportion to the degree of risk. In the aftermath of such an event, there can be a heightened sense of alarm, which triggers a desire to do whatever is required to prevent a repeat, no matter how extreme or costly. A lesson of Christchurch Cathedral is that whatever the precautions, a set of circumstances can render them ineffective. On balance, therefore, it seems reasonable to retain the status quo on older buildings, and insist on earthquake strengthening only when they are being modified. It would, however, be very useful if homeowners were advised individually how earthquake-resistant their houses were. They could then decide whether to strengthen or sit tight. It would also be helpful, as the United Future leader, Peter Dunne, suggests, if earthquake-prone buildings were publicly listed. People should know the status of buildings they live in, work in or use often.

Unfortunately – as the Tohoku sea walls showed – mitigation policies are often developed without careful consideration of their benefits and costs. Communities are often unclear about what they are buying and how much they are paying. Because they are playing against nature without a clear strategy, it is not surprising that they sometimes do badly. Doing better requires selecting strategies to best use their limited resources. This is not easy, because the benefits of various strategies cannot be estimated precisely, given our limited ability to estimate the occurrence and effects of future events. However, even simple estimates of the costs and benefits of different strategies often show that some make much more sense than others.

Table 1.1   US deaths from various causes in 1996

Cause of death	No. of deaths
Heart attack	733,834
Cancer	544,278
Motor vehicles	43,300
AIDS	32,655
Suicide	30,862
Liver disease/Cirrhosis	25,135
Homicide	20,738
Falls	14,100
Poison (accidents)	10,400
Drowning	3,900
Fire	3,200
Bicycle accidents	695
Severe weather	514
Animals	191
In-line skating	25
Football	18
Skateboards	10

Stein and Wysession, 2003. Reproduced with permission of John Wiley & Sons.

A key point in allocating resources is that natural hazards are only one of the many problems society faces. Comparing the numbers of deaths per year in the US from various causes (Table 1.1) brings out this point.

US earthquakes have caused an average of about twenty deaths per year since 1812. The precise number depends mostly on how many died in the 1906 earthquake that destroyed much of San Francisco, which is not known well. This analysis starts after 1812 because it is not known if anyone was killed in that year's big (about magnitude 7) New Madrid earthquakes in the Midwest.

These numbers vary from year to year and have uncertainties because of the way they are reported, but give useful insights into risks. For example, because there are about 300 million people in the US, the odds of being killed by an animal are about 200/300,000,000 or 1 in 1.5 million.

As you would expect, these numbers show that earthquakes are not a major cause of deaths in the US Although earthquakes are dramatic and can cause major problems, many more deaths result from causes like drowning or fires. Severe weather is about 25 times more dangerous than earthquakes. Earthquakes rank at the level of in-line skating or football, and severe weather is

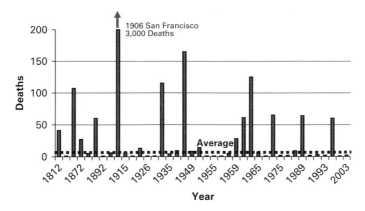

Figure 1.7 Annual deaths in the United States from earthquakes, 1812–2003.

at the level of bicycle accidents. Hence during the 1950s and 1960s seismologist Charles Richter was an early advocate for earthquake-resistant construction in California while pointing out, "I don't know why people in California or anywhere worry so much about earthquakes. They are such a small hazard compared to things like traffic."

This relatively low earthquake danger arises because most US earthquakes do little harm. Even those felt in populated areas are commonly more of a nuisance than a catastrophe. In most years, no one is killed by an earthquake (Figure 1.7). About every 40 years an earthquake kills more than 100 people, and the 1906 San Francisco earthquake is thought to have killed about 3000 people. This pattern arises because big earthquakes are much less common than small ones, and large numbers of deaths occur when a rare big earthquake takes place where many people live, such as happened with the San Francisco earthquake. Other natural disasters like hurricanes behave the same way, with rare large events doing the most damage. Because people remember dramatic events and do not think about how rare they are, it is easy to forget that more common hazards are much more dangerous.

Earthquakes can also cause major property damage. Although the 1994 Northridge earthquake was not that big – magnitude 6.7 – it happened under the heavily populated Los Angeles metropolitan area, and caused 58 deaths and $20 billion in property damage. Still, this damage is only equivalent to about 10% of the US annual loss due to automobile accidents.

Earthquakes are only a secondary hazard in the US because large earthquakes are relatively rare in heavily populated areas, and buildings in the most active areas such as California are built to reduce earthquake damage. Earthquakes are a bigger problem in some other countries where many people live near plate boundaries. Although the statistics are sometimes imprecise, major earthquakes can be very destructive, as the Tohoku earthquake shows. The highest property losses occur in developed nations where more property is at risk, whereas fatalities are highest in developing nations.

Over the past century, earthquakes worldwide have caused about 11,500 deaths per year. This number is increasing over time as populations at risk grow, and averaged about 20,000 deaths per year from 2000–2009. Still, it is a much lower figure than the number of deaths caused by diseases. For example, AIDS and malaria cause about 1.8 million and 655,000 deaths per year, respectively.

Similarly, other natural hazards cause infrequent, but occasionally major, disasters involving higher numbers of fatalities and greater damage. As a result, society needs to think carefully about what to do. We want to mitigate natural hazards, but not focus on them to the extent that we unduly divert resources from other needs.

1.3   How Can We Do Better?

A frequent limitation of current approaches is that of treating the relevant geoscience, engineering, economics, and policy formulation separately. Geoscientists generally focus on using science to assess hazards; engineers and planners focus on mitigation approaches; and economists focus on costs and benefits. Each group often focuses on its aspect of the problem, does not fully appreciate how the others think, what they know, and what they do not.

This situation often leads to policies that make little scientific or economic sense. Hazard assessments often underestimate the limits of scientific knowledge. Mitigation policies are often developed without considering their costs and benefits. The net result is that communities often overprepare for some hazards and underprepare for others.

For example, since 1978 the Japanese government has followed a law called the Large-Scale Earthquake Countermeasures Act that requires operating a monitoring system to detect precursors – i.e., changes in properties of the earth – which are supposed to allow a large earthquake along part of the Japan Trench (Figure 1.3) to be predicted. In theory what should happen is that a panel of five geophysicists will review the data and determine that a large earthquake is imminent, the director of the Japan Meterological Agency will inform the prime minister, and the cabinet will then declare a state of emergency, which will stop almost all activity in the nearby area. The problem is that, as we will discuss, such precursors have never been reliably observed, so at present there is no way to accurately predict earthquakes.

Another good example is the way the US government treats different hazards. It wants buildings built for the maximum wind speed expected on average once every 50 years, the typical life of a building, which there's a 2% (1/50) chance of having in any one year. However, it tells communities to plan for the maximum flooding expected on average once every 100 years, or that there's a 1% chance of having in any one year. It wants even higher standards for earthquakes. California should plan for the maximum shaking expected on average once in 500 years, and Midwestern states for the maximum shaking expected on average once in 2500 years. This pattern is the opposite of what one might expect, because wind and flooding – often due to the same storm – cause much more damage than earthquakes. None of these time periods come from careful analysis, and it is not clear which if any should be different. It might better to prepare a 500-year plan for both floods and earthquakes. We will see that using 2500 years is likely to over-prepare for earthquakes. Conversely, it seems that in many areas planning only for the 100-year flood gives too low a level of protection, so it would be wise to prepare for larger floods.

This book explores ways of taking a broader view that can help in developing more sensible policies. Policy-making can be viewed as the intersection of the different approaches (Figure 1.8).

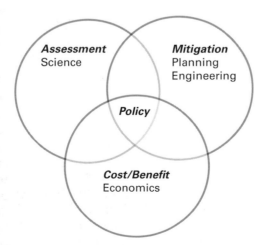

Figure 1.8   Schematic illustrating how formulating hazard policy involves integrating assessment, mitigation, and economics.

One attempt to take a broad view came from the Copenhagen Consensus, a group that evaluated ways to spend an additional $75 billion worldwide (about 15% of global aid spending) to best improve public health. Their top priorities came out as:

1. Nutrition supplements
2. Malaria treatment
3. Childhood immunization
4. Deworming school children
5. Tuberculosis treatment
6. Research to enhance crop yields
7. Natural hazard warning systems
8. Improving surgery
9. Hepatitis B immunization
10. Low cost drugs to prevent heart attacks

Hence in their view natural hazards emerge as a major, but not absolute top, priority.

More effective natural hazards policy can be developed both by advancing each of the relevant disciplines and integrating their knowledge and methods. Fortunately there is an increasing awareness of the need for both, especially among young researchers who would like to do a better job of mitigating hazards.

This book is an overview of some aspects of this challenge. It is written assuming readers have diverse backgrounds in geoscience, engineering, economics, and policy studies. We use the Tohoku earthquake to illustrate some key issues, and then introduce some basic concepts to help readers appreciate the value of the other disciplines and their interrelations, and develop the background to explore more advanced treatments of these topics. We explore aspects of what we know, what we do not know, what mitigation approaches are available, and how we can choose between them. Although we primarily use earthquakes and tsunamis as examples, most of the discussion in this book applies to other natural hazards.

Beyond the scientific and economic issues, the mitigation policies a community chooses reflect sociocultural factors. Because societies overrate some risks and underrate others, what they will spend to mitigate them often has little relation to the actual risk. The policies chosen also reflect interest groups and political influence. For example, the billions of dollars spent on seawalls in Japan is a cost to the nation as a whole, but a benefit to politically connected contractors. In North Carolina, political policy prohibits state coastal planning officials considering the possibility that the warming climate will accelerate the rate of sea level rise, as anticipated from melting ice caps. These factors are beyond our scope here.

For these and other reasons, neither we nor anyone else can offer the "right" way of solving these problems, because no unique or right answers exist for a particular community, much less all communities. However, the approaches we discuss can help communities make more informed and better decisions.

Our view of how to use science in a careful and open process for formulating policy is summarized in an eloquent statement by Nobel Prize winning physicist Richard Feynman. In 1986, after the explosion of the space shuttle *Challenger*, the US government followed its usual practice of appointing a nominally "independent" commission to study the accident. The commissioners who, other than Feynman, were insiders from NASA, government, and the aerospace industry, wanted to support NASA rather than ask hard questions. Their supportive report was instantly forgotten. However, Feynman's dissenting assessment, explaining what went wrong and why, become a classic example of objective outside analysis. It is remembered especially for its conclusion:

> NASA owes it to the citizens from whom it asks support to be frank, honest, and informative, so these citizens can make the wisest decisions for the use of their limited resources. For a successful technology, reality must take precedence over public relations, for nature cannot be fooled.

Questions

1.1. Although the losses from natural disasters are very large, they can be viewed in various ways. How large are these losses per person on earth? How do they compare to the world's total military budget?

1.2. Comparing California and Alaska, which would you expect to have the higher earthquake hazard? Which should have the higher risk?

1.3. Of the approximately 50 people killed each year in the US by lightning, about 80% are male. Analyze the difference between male and female deaths in terms of hazard, vulnerability, and risk. Suggest possible causes for the difference and how to test these hypotheses. For example, what would your hypotheses predict for the geographic distribution?

1.4. In thinking about hazards, it is useful to get a sense of the order of magnitudes involved. This approach is sometimes called "Fermi estimation" after Nobel Prize winning physicist Enrico Fermi, who used to ask students in qualifying exams questions like "How many piano tuners are in Chicago?" Estimate the order of magnitude – 1, 10, 100, or 1000 – of the number of deaths per year in the US caused by bears, sharks, bees, snakes, deer, horses, and dogs. A good way to tackle this is to put them in the relative order you expect, and then try to estimate numerical values.

1.5.   A useful way to get insight into risks is to compare them. For example, because about 700 people a year in the US are killed in bicycle accidents and the US population is about 300 million people, the odds of being killed in a bicycle accident are about 700/300,000,000 or 1 in about 430,000. What are the odds of being killed in an earthquake in the US? Because such estimates depend somewhat on whether a disastrous earthquake has occurred recently, estimate the odds both with the data given in this chapter and assuming another earthquake as disastrous as the 1906 one had occurred recently.

1.6.   An interesting comparison with natural hazards is the odds of winning a state lottery. If the lottery involves matching six numbers drawn from a field of 1 to 49, the odds can be found as follows: The first number drawn can be any of the 49 numbers, and you need one of the six numbers on your ticket to match it. The second number is one of the 48 numbers left, and you need one of the remaining five numbers on your ticket to match it, and so on. Thus your odds of winning are $6/49 \times 5/48 \times 4/47 \times 3/46 \times 2/45 \times 1/44$. Calculate this number and compare it to the odds of being killed in an earthquake or a bicycle accident.

1.7.   The US's National Center for Missing and Exploited Children has said, "Every day 2000 children are reported missing." Use this number to estimate the fraction of the nation's children that would be missing each year. How does this prediction compare to your experience during the years you spent in school? From your experience, estimate a realistic upper bound for this fraction. How might the much larger 2000 per day number have arisen?

1.8.   The enormous destruction to New Orleans by hurricane Katrina in 2005 had been predicted for years, because about half of the city is below sea level and human actions caused land subsidence along the coast that increased the destructive power of hurricanes. Some argued that the city should not be rebuilt at its present site, because it would be at risk of a similar disaster. Others argued that the site is too important culturally, economically, and historically to abandon, and that it could be made safe. How would you decide between options of not rebuilding the coastal defenses that failed, rebuilding them to deal with a similar storm, or building ones to deal with larger storms?

1.9.   How do you respond to the *New Zealand Herald* editorial quoted in section 1.2? Do you agree or disagree, and why?

1.10.   Almost 2000 years ago, Pompeii and other cities near present Naples, Italy were destroyed by an eruption of the volcano Vesuvius. Since the last eruption in 1944, the Bay of Naples region has been a hotbed of construction – much of it unplanned and illegal – that has hugely increased the number of people living in the danger zone of the volcano. Millions of people may be affected by the next eruption, with those in the "red zone" (zona rossa) under the most serious threat. The authorities are considering paying these people to relocate. How would you formulate and evaluate such plans?

1.11.   What do you consider to be the five major problems facing your community? Which, if any, involve natural hazards?

Further Reading and Sources

Kieffer (2013) gives an overview of natural disaster science.

Figure 1.1 is taken from "Natural catastrophes 2012" available at *https://www.munichre.com/touch/naturalhazards/en/homepage/default.aspx*

The *Economist* (January 14, 2012; "Counting the cost of calamities", *http://www.economist.com/node/21542755/print*) reviews the cost of natural disasters. Global fatality data are given in Guha-Sapir et al. (2012). Natural disaster loss and cost issues are discussed by Kliesen (1994) and National Research Council, Committee on Assessing the Costs of Natural Disasters (1999).

Thorp's (1966) book presented the winning strategies for blackjack (see also *http://www.edwardothorp.com*). Fictionalized accounts of the MIT blackjack teams given by Mezrich (2003; *http://www.youtube.com/watch?v=QflVqavHHM0*) are the basis of the film "21."

Fritz et al. (2012) discuss the Tohoku tsunami's effects in the area of Figure 1.1. Videos of the tsunami are linked at *http://www.geologyinmotion.com/2011/03/more-videos-of-tsunami-and-situation-in.html*. The resulting losses are summarized by Normile (2012).

Cyranoski (2012a), Harner (2012), and Tabuchi (2012) discuss tsunami policy for the Nankai area. The "worst case" comment is from Harner (2012). O'Connor (2012) reviews the number of earthquake deaths in Haiti. Peresan and Panza (2012) discuss the history of the Italian earthquake hazard map.

The *New Zealand Herald* editorial is dated March 4, 2011. Table 1.1 is from Stein and Wysession (2003). Data in Figure 1.5 are from *http://earthquake .usgs.gov/regional/states/us_deaths.php*. The Richter quotation is from Hough's (2007) biography. Geller (2011) describes the Japanese government prediction policy. The Copenhagen Consensus priorities are from *http://www .copenhagenconsensus.com/projects/copenhagen-consensus-2012/outcome*.

Lee (2012) summarizes the North Carolina sea level issue. Feynman's activities on the Challenger commission are described by Gleick (1992) and his dissent to the commission's report is reprinted in Feynman (1988) and available at *http://science.ksc.nasa.gov/shuttle/missions/51-l/docs/rogers -commission/Appendix-F.txt*.

Weinstein and Adam (2008) explain Fermi estimation with examples including the total length of all the pickles consumed in the US in one year. Ropeik and Gray (2002) and Aldersey-Williams and Briscoe (2008) give general audience discussions of risks. The US National Incidence Studies of Missing, Abducted, Runaway, and Thrownaway Children (Finkelhor et al., 2002; *https://www.ncjrs.gov/html/ojjdp/nismart/03/*) found that in 1999 there were an estimated 115 stereotypical kidnappings, defined as abductions per-petrated by a stranger or slight acquaintance and involving a child who was transported 50 or more miles, detained overnight, held for ransom or with the intent to keep the child permanently, or killed.

References

Aldersey-Williams, H., and S. Briscoe, *Panicology: Two Statisticians Explain What's Worth Worrying About (and What's Not) in the 21st Century*, Penguin, New York, 2008.

Cyranoski, D., Tsunami simulations scare Japan, *Nature*, *484*, 296–297, 2012a.

Cyranoski, D., Rebuilding Japan, *Nature*, *483*, 141–143, 2012b.

Feynman, R. P., *What Do You Care What Other People Think*, W. W. Norton, New York, 1988.

Finkelhor, D., H. Hammer, and A. Sedlak, Nonfamily abducted children: national estimates and characteristics, U.S. Department of Justice, 2002.

Fritz, H. M., D. A. Phillips, A. Okayasu, T. Shimozono, H. Liu, F. Mohammed, V. Skanavis, C. E. Synolakis, and T. Takahashi, The 2011 Japan tsunami current veloc-ity measurements from survivor videos at Kesennuma Bay using LiDAR, *Geophys. Res. Lett.*, *39*, L00G23, doi: 10.1029/2011GL050686, 2012.

Geller, R. J., Shake-up time for Japanese seismology, *Nature*, *472*, 407–409, 2011.

Gleick, J., *Genius: The Life and Science of Richard Feynman*, Pantheon, New York, 1992.

Guha-Sapir, D., F. Vos, R. Below, and S. Ponserre, *Annual Disaster Statistical Review 2011*, Centre for Research on the Epidemiology of Disasters, Brussels, 2012. (http://www.cred.be/sites/default/files/ADSR_2011.pdf).

Harner, S., BTW, get ready for a 34 meter tsunami, *Forbes*, April 2, 2012.

Hough, S. E., *Richter's Scale: Measure of an Earthquake, Measure of a Man*, Princeton University Press, Princeton, NJ, 2007.

Kieffer, S. W., *The Dynamics of Disaster*, W.W. Norton, New York, 2013.

Kliesen, K., *The Economics of Natural Disasters*, Federal Reserve Bank of Saint Louis, 1994. (http://www.stlouisfed.org/publications/re/articles/?id=1880).

Lee, J. J., Revised North Carolina sea level rise bill goes to governor, *Science Insider*, July 3, 2012.

Lomnitz, C., Comment on "temporal and magnitude dependance in earthquake recurrence models" by C. A. Cornell and S. R. Winterstein, *Bull. Seismol. Soc. Am.*, 79, 1662, 1989.

Mezrich, B., *Bringing Down the House: The Inside Story of Six MIT Students Who Took Vegas for Millions*, Free Press, Old Tappan, NJ, 2003.

National Research Council, Committee on Assessing the Costs of Natural Disasters, *The Impacts of Natural Disasters: A Framework for Loss Estimation*, 1999. (http://www.nap.edu/catalog/6425.html).

Normile, D., One year after the devastation, Tohoku designs its renewal, *Science*, 335, 1164–1166, 2012.

O'Connor, M., Two years later, Haitian earthquake death toll in dispute, *Columbia Journalism Review*, January 12, 2012.

Peresan, A., and G. Panza, Improving earthquake hazard assessments in Italy: an alternative to Texas sharpshooting, *Eos Trans. AGU*, 93, 538, 2012.

Ropeik, D., and G. Gray, *Risk: A Practical Guide for Deciding What's Really Safe and What's Really Dangerous in the World Around You*, Houghton Mifflin, New York, 2002.

Stein, S., *Disaster Deferred: How New Science is Changing our View of Earthquake Hazards in the Midwest*, Columbia University Press, New York, 2010.

Stein, S., and M. Wysession, *Introduction to Seismology, Earthquakes, and Earth Structure*, Blackwell, Oxford, 2003.

Stein, S., R. J. Geller, and M. Liu, Why earthquake hazard maps often fail and what to do about it, *Tectonophysics*, 562–563, 623–626, 2012.

Stein, S., R. J. Geller, and M. Liu, Reply to comment by A. Frankel on "Why earthquake hazard maps often fail and what to do about it", *Tectonophysics*, 592, 207–209, 2013.

Tabuchi, H., Tsunami projections offer bleak fate for many Japanese towns, *New York Times*, April 9, 2012.

Thorp, E., *Beat the Dealer: A Winning Strategy for the Game of Twenty-One*, Knopf, New York, 1966.

Weinstein, L., and J. Adam, *Guesstimation: Solving the World's Problems on the Back of a Cocktail Napkin*, Princeton University Press, Princeton, NJ, 2008.

2

When Nature Won

It was a profoundly man-made disaster – that could and should have been foreseen and prevented.

Fukushima Nuclear Accident Independent Investigation
Commission (2012)[1]

2.1 The Best-Laid Plans

Until March 11, 2011, residents of Japan's Tohoku coast were proud of their tsunami defenses. The 10-meter high sea walls that extended along a third of the nation's coastline – longer than the Great Wall of China – cost billions of dollars and cut off ocean views. However, these costs were considered a small price to pay for eliminating the threat that had cost many lives over the past hundreds of years. In the town of Taro, people rode bicycles, walked, and jogged on top of the impressive wall. A school principal explained, "For us, the sea wall was an asset, something we believed in. We felt protected."

The earthquake and tsunami hazard result from Japan's location, where four plates interact. Historically, the largest hazard arose from earthquakes on the boundaries to the east (Figure 2.1), where the Pacific plate subducts at the Japan Trench and the Philippine Sea plate subducts at the Nankai Trough.

Beneath Tohoku, the west-dipping Pacific plate can be identified by earthquakes that occur within it, down to a depth of about 200 km. However, most

[1] Fukushima Nuclear Accident Independent Investigation Commission, 2012.

Playing against Nature: Integrating Science and Economics to Mitigate Natural Hazards in an Uncertain World, First Edition. Seth Stein and Jerome Stein.
© 2014 John Wiley & Sons, Ltd. Published 2014 by John Wiley & Sons, Ltd.
Companion Website: www.wiley.com/go/stein/nature

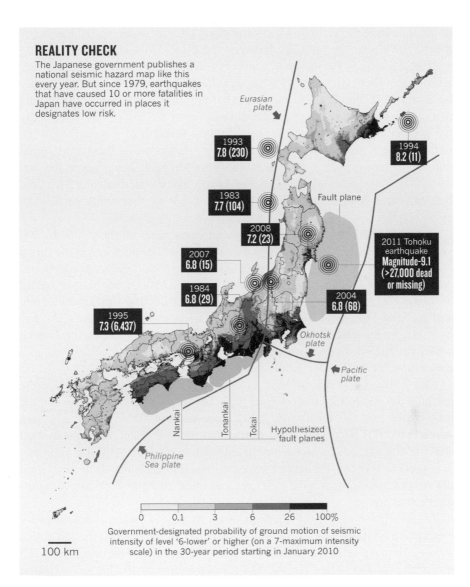

REALITY CHECK

The Japanese government publishes a national seismic hazard map like this every year. But since 1979, earthquakes that have caused 10 or more fatalities in Japan have occurred in places it designates low risk.

Eurasian plate

1993
7.8 (230)

1994
8.2 (11)

1983
7.7 (104)

Fault plane

2008
7.2 (23)

2011 Tohoku earthquake
Magnitude-9.1
(>27,000 dead or missing)

2007
6.8 (15)

1984
6.8 (29)

2004
6.8 (68)

1995
7.3 (6,437)

Okhotsk plate

Pacific plate

Nankai   Tonankai   Tokai

Hypothesized fault planes

Philippine Sea plate

0   0.1   3   6   26   100%

Government-designated probability of ground motion of seismic intensity of level '6-lower' or higher (on a 7-maximum intensity scale) in the 30-year period starting in January 2010

100 km

Figure 2.1   Comparison of Japanese government hazard map to the locations of earthquakes since 1979 that caused 10 or more fatalities. Hazard is shown as probability that the maximum ground acceleration (shaking) in any area would exceed a particular value during the next 30 years. Larger expected shaking corresponds to higher predicted hazard. The Tohoku area is shown as having significantly lower hazard than other parts of Japan, notably areas to the south. Since 1979, earthquakes that caused 10 or more fatalities occurred in places assigned a relatively low hazard. (Geller, 2011. Reproduced with permission of *Nature*.) See also color plate 2.1.

of the time little seems to be happening along the great fault, called the mega-thrust, that forms the interface between the plates. In reality, a lot is going on. Every year, the Pacific plate converges on Japan at about 80 mm per year and subducts beneath Japan. However, the megathrust fault is locked, so strain builds up on it (Figure 2.2). This strain buildup causes motions of the ground that can be measured using the Global Positioning System (GPS). Eventually the accumulated strain exceeds the frictional strength of the fault, and the fault slips in a great earthquake, generating seismic waves that can do great damage. The overriding plate that had been dragged down since the last earthquake rebounds and displaces a great volume of water, causing a tsunami that can have devastating effects. Over hundreds of years, this earthquake cycle repeats.

A simple estimate shows how large an earthquake might result. In 100 years, $100 \times 80 = 8000$ mm or 8 m of motion might accumulate at the thrust fault. If all this were released at once, a very large earthquake would occur. As a result, the Japanese hazard planners used the historic earthquake record to divide the trench into segments of the trench about 100–150 km in length, and inferred how large an earthquake to expect on each (Figure 2.3). For example, for the area of the Japan Trench off Miyagi prefecture on the Tohoku coast, the hazard mappers assumed that there was a 99% probability that a magnitude 7.5 earthquake would occur in the next 30 years.

This forecast and similar ones for other regions were used to produce the national seismic hazard map (Figure 2.1). The map showed predictions of the probability that the maximum ground acceleration (shaking) in any area would exceed a particular value during the next 30 years. Larger expected shaking corresponds to higher predicted seismic hazard. A similar approach was used to forecast the largest expected tsunami. Engineers, in turn, used the results to design tsunami defenses and build structures to survive earth-quake shaking.

All this planning proved inadequate on March 11, when a magnitude 9 earthquake offshore generated a huge tsunami that overtopped the sea walls. This earthquake released about 150 times the energy of the magnitude 7.5 quake that was expected for the Miyagi-oki region by the hazard mappers. Somehow, the mapping process significantly underpredicted the earthquake hazard.

2.2   Why Hazard Assessment Went Wrong

The hazard map predicted less than a 0.1% probability of shaking with inten-sity "6-lower" on the Japan Meteorological Agency intensity scale in the next

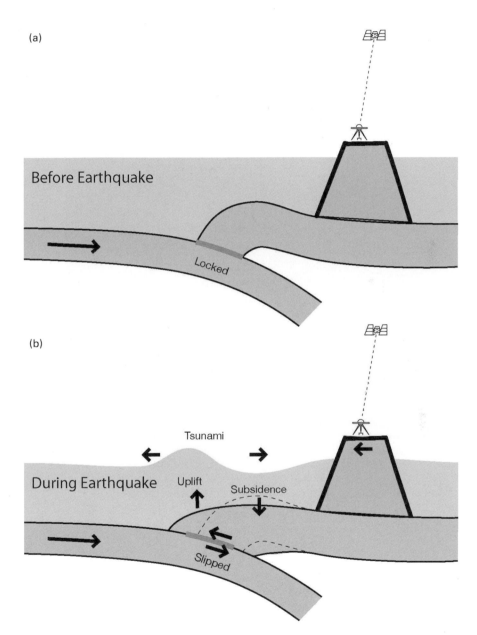

Figure 2.2   Schematic illustration of the earthquake cycle on the locked megathrust fault at a subduction zone. Strain builds up for many years (a) until it is released in a major earthquake (b). The strain accumulation can be measured using techniques including GPS.

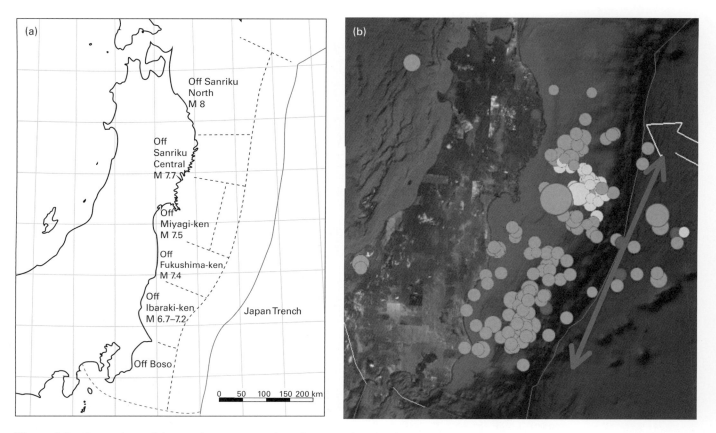

Figure 2.3   Comparison of the trench segments and maximum earthquake sizes assumed in the Japanese hazard map (a) to the aftershock zone of the March 11 earthquake (b), which broke five segments. (Stein et al., 2012. Reproduced with permission of Elsevier B.V.)

30 years off Tohoku. In other words, such shaking was expected on average only once in the next 30/0.001 or 30,000 years. However, within two years, such shaking occurred.

The question that arose immediately was whether this situation indicated a bad map or simply bad luck. Some argued that this was a rare and unpredictable event that should not be used to judge the map as unsuccessful. Such events are termed "black swans" because before Europeans reached Australia, all swans were thought to be white. However, the failure turned out to reflect deficiencies in the hazard assessment for Tohoku. Moreover, it turns out that similar problems have arisen elsewhere, indicating systemic difficulties in earthquake hazard mapping.

In fact, Tohoku illustrated four partially overlapping factors that can cause a earthquake hazard map to fail:

- *Bad physics*: incorrect physical models of the faulting processes;
- *Bad assumptions*: about which faults are present, which are active, how fast they are accumulating strain, and how it will be released;
- *Bad data*: data that are "wrong" in the sense that they are lacking, incomplete, or misinterpreted;
- *Bad luck*: because the map gives probabilities, rare events can produce higher shaking without invalidating the map.

Bad physics caused the widespread view among Japanese seismologists that giant magnitude 9 earthquakes would not occur on the Japan Trench off Tohoku. Thus the largest future earthquakes along different segments of the trench were expected to have magnitudes between 7 and 8 (Figure 2.3). This view led to the incorrect assumption that different segments of the trench would not break simultaneously, so the largest earthquake on any would have magnitude 8.

As illustrated in Figure 2.4a, a magnitude 9 earthquake involves a larger average slip over a larger fault area, resulting in a larger tsunami because the maximum tsunami run-up height is typically about twice the fault slip. The March earthquake broke five segments of the trench, giving a magnitude 9 earthquake and a huge tsunami that overtopped 10-meter high sea walls.

Such a giant earthquake was not anticipated due to several incorrect assumptions that reinforced one another. First, the earthquake history from seismology, spanning the roughly 100 years since the invention of the seismometer in the 1880s, appeared to show no record of such giant earthquakes. However, in the decade prior to 2011, increasing attention was being paid to geological and historical data showing that large tsunamis had struck the area in 869, 1896, and 1933. Some villages had stone tablets marking the heights

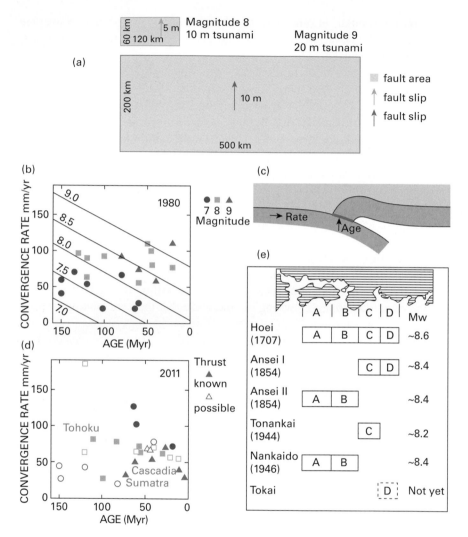

Figure 2.4 What went wrong at Tohoku. (a) Illustration of the relative fault dimensions, average fault slip, and average tsunami run-up for magnitude 8 and 9 earthquakes. (b) Data available in 1980, showing the largest earthquake known at various subduction zones. Magnitude 9 earthquakes had been observed only where young lithosphere subducts rapidly. Diagonal lines show predicted maximum earthquake magnitude. (c) Physical interpretation of this result in terms of strong mechanical coupling and thus large earthquakes at the trench interface. (d) Update of (b) with data including the 2004 Sumatra and 2011 Tohoku earthquakes. (e) Earthquake history for the Nankai trough area illustrating how different segments rupturing cause earthquakes of different magnitudes. Segment "D" is the presumed Tokai seismic gap. (Stein and Okal, 2011. Reproduced with permission of American Geophysical Union.) See also color plate 2.4.

reached by previous tsunamis and warning "Do not build your homes below this point."

Second, the earthquake record seemed consistent with the hypothesis that the physics of subduction precluded M 9 earthquakes on the Japan Trench. In 1980, an analysis of the largest known earthquakes at various subduction zones (Figure 2.4b) showed a striking pattern – magnitude 9 earthquakes had only occurred where portions of a plate younger than 80 million years old was subducting rapidly, faster than 50 mm/yr. This result made intuitive sense, because both young age and faster convergence could favor strong mechanical coupling at the interface between the two plates (Figure 2.4c). This model was widely accepted until December 26, 2004, when the giant magnitude 9.3 Sumatra earthquake generated the devastating Indian Ocean tsunami. According to the model, this trench should have generated at most an earthquake with a magnitude of about 8. However, with newer data including Sumatra the apparent correlation vanished, as the 2011 Tohoku earthquake confirmed (Figure 2.4d). Thus instead of only some subduction zones being able to generate magnitude 9 earthquakes, it now looks like many or all can. The apparent pattern had resulted from the small sample. Magnitude 9 earthquakes are rare – on average fewer than one per decade – making them about ten times rarer than magnitude 8 earthquakes. As we will discuss, our ideas about where large earthquakes can occur are often biased by the short record available.

Third, although the hazard map assumed that only one segment would rupture at a time, five actually broke. The magnitude of earthquakes depends on their seismic moment, which is the product of the area of fault that ruptured, the average distance the ground in the ruptured area slipped, and the shear modulus – a measure of the strength of the rock. The ruptured area has the largest effect, because subduction zone earthquakes break various segments of a trench, as shown in Figure 2.4e for the Nankai Trough. Sometimes one segment ruptures, and other times more than one does. The more segments rupture, the bigger the earthquake, as the 1707 earthquake shows.

Fourth, the presumed absence of giant earthquakes on the Japan Trench was implicitly interpreted as indicating that much of the subduction occurred aseismically (i.e. without earthquake activity), so most of the plate motion would not show up as earthquakes. The Kurile trench, just to the north, seemed to show this discrepancy. The largest seismologically recorded earthquakes there are magnitude 8, which only account for about one third of the plate motion. Hence it had been assumed that most of the subduction occurred aseismically. However, recently discovered deposits from ancient tsunamis show that much larger earthquakes had happened in the past, accounting for much of the subduction that had been thought to occur aseismically. In hindsight, the same applied off Tohoku. GPS data, which were becoming available, showed a much higher rate of strain accumulation on the plate interface than would be expected if a large fraction of the subduction occurred aseismically. Including these data in the hazard assessment would have strengthened the case for considering the possibility of large earthquakes.

However, the revised ideas about maximum earthquake and tsunami size were not yet appreciated and incorporated into the hazard map. Thus, as summarized by Sagiya (2011),

> If historical records had been more complete, and if discrepancies between data had been picked up, we might have been alert to the danger of a magnitude-9 earthquake hitting Tohoku, even though such an event was not foreseen by the Japanese government.

2.3   How Mitigation Fared

The iconic images of the tsunami pouring over 5–10 m high sea walls illustrated the limitations of these mitigation structures. Of the region's 300 km of seawalls, 180 were washed away (Figure 2.5). The extensive damage

Figure 2.5   Seawall destroyed by the tsunami. (Courtesy of Hermann M. Fritz.)

occurred because the tsunami was much larger than those from the earth-quakes in 1896 and 1933. As a result, a much larger area was flooded, and buildings that survived the earlier tsunamis were destroyed. A coastal forest that had been an effective breakwater against the earlier tsunamis was washed away.

The most crucial mitigation failure resulted in the destruction of the Fuku-shima nuclear power plant. The process used to design safety features under-predicted the hazard, leading to inadequate safety features, notably a seawall much lower than tsunami waves that could have been reasonably expected at the time the power plant was built in the 1960s. Although this vulnerability arose because of limited knowledge at that time, both the plant operator and government regulators ignored warnings of the higher hazard showed by new scientific results. Further vulnerability resulted from the fact that the emer-gency generators were located at the plant and so flooded and became inoperable.

Measures with only moderate cost that could have reduced these vulner-abilities, including putting the emergency generators on higher ground, were not taken, because of institutional factors. In particular, the combination of government and industry promoted the pervasive "safety myth" criticized by in Kazuyoshi Saito's protest song:

> If you walk across this country, you'll find 54 nuclear reactors
> School textbooks and commercials told us they were safe.
> It was always a lie, it's been exposed after all
> It was really a lie that nuclear power is safe.

The Japanese commission that investigated the power plant disaster con-cluded that

> the subsequent accident at the Fukushima Daiichi nuclear power plant cannot be regarded as a natural disaster. It was a profoundly man-made disaster – that could and should have been foreseen and prevented. Our report catalogues a multitude of errors and willful negligence that left the Fukushima plant unpre-pared for the events of March 11. . . . What must be admitted – very painfully – is this was a disaster 'Made in Japan'. Its fundamental causes are to be found in the ingrained conventions of Japanese culture: our reflexive obedience, our reluctance to question authority, our devotion to 'sticking with the program', our groupism, and our insularity.

While recognizing national differences, our view is these problems reflect the general nature of large bureaucracies and vested interests, which are similar in different countries. Analogous organizational failures elsewhere have been major contributors to disasters. These include the 1986 loss of the US space shuttle *Challenger*, the destruction of much of New Orleans by Hurricane Katrina in 2005, and the 2008 collapse of the US housing and financial markets.

Other mitigation policies were much more successful. Immediately following the earthquake, the tsunami warning system issued alerts. An extensive education program including tsunami evacuation drills had been conducted over the years. Many residents evacuated to higher ground, based on the warning and the fact that the shaking was longer and stronger than they had experienced in their lives. Unfortunately, the first warnings underestimated the size of the tsunami, inducing some people not to evacuate due to their confidence in the breakwaters, and some sites designated as tsunami shelters were flooded by the larger-than-planned-for tsunami.

Building codes for earthquake-resistant construction were also generally effective. Structural damage from ground shaking was generally minor, due both to appropriate construction and to the fact that the earthquake occurred far offshore, because the strength of ground shaking decreases rapidly with distance from an earthquake.

2.4   The Challenges Ahead

The fact that such enormous damage occurred in an affluent nation that had devoted major resources to earthquake and tsunami hazard assessment and mitigation demonstrates the limitations of current strategies, and the need to improve them. Although these problems were starting to be recognized, they were not fully appreciated. Within the scientific community, this is starting to change after the Tohoku earthquake.

The first issue is what went wrong for Tohoku in particular. The second issue is whether this failure was an exceptional case, or whether it indicates systemic difficulties that arise in earthquake and tsunami hazard mapping and mitigation. The third issue is how to improve this situation. This process is similar to what occurs after an airplane crash, in which investigators first try to find what went wrong, then to learn whether the faults occurred because of special circumstances or reflected a deeper problem, and finally explore solutions.

The case that the Tohoku failure reflected a systemic problem in hazard asessment was made by Geller (2011). He noted that the map that underestimated the Tohoku hazard (Figure 2.1) instead focused on the Nankai Trough area, due to the internal politics of government and university organizations involved with hazard assessment. Based on the concept of "seismic gaps," in which areas that have not ruptured for long enough are "due" to break, large earthquakes are expected on the Nankai (segments A-B), Tonankai (segment C) and especially Tokai (segment D) portions of the trench (Figure 2.4). Geller points out that:

> The regions assessed as most dangerous are the zones of three hypothetical 'scenario earthquakes' (Tokai, Tonankai and Nankai; see map). However, since 1979, earthquakes that caused 10 or more fatalities in Japan actually occurred in places assigned a relatively low probability. This discrepancy – the latest in a string of negative results for the characteristic earthquake model and its cousin, the seismic-gap model – strongly suggests that the hazard map and the methods used to produce it are flawed and should be discarded.

This analysis pointed out that the map seemed to be of less use than it would be by assuming that the earthquake hazard was the same everywhere in Japan. Fundamentally, the map underestimated how variable earthquakes are in space and time. The problem was its assumptions, which we will discuss later, that all future large earthquakes will be like those known to have happened and are more likely to occur on parts of the subduction zone where none have occurred recently.

The hazard map failure for Tohoku was not unusual. Other highly destructive earthquakes such as the 2010 Haiti and 2008 Wenchuan, China, earthquakes have occurred in areas predicted by hazard maps to be relatively safe. As a result, large earthquakes cause many more fatalities worldwide than would have been expected from hazard maps. *Science* magazine described this situation as the

> seismic crystal ball proving mostly cloudy around the world. . . . In China, New Zealand, and California as well, recent earthquakes have underscored scientists' problems forecasting the future. A surprisingly big quake arrives where smaller ones were expected, as in Japan; an unseen fault breaks far from obviously dangerous faults, as in New Zealand. And, most disconcerting, after more than 2 decades of official forecasting, geoscientists still don't know how much confidence to place in their own warnings.

Thus after the Tohoku earthquake, previously raised questions about the mapping methods are receiving greater attention.

Many traditional mitigation approaches are also being reexamined. Because rebuilding the coastal defenses to withstand tsunamis as large as that which struck Tohoku is too expensive, those planned are about 12 m high, only a few meters higher than the previous defenses. These are planned to provide protection for the largest tsunamis expected every 200–300 years, augmented with land-use planning to provide some protection against much larger tsunamis. The defenses should reduce economic losses, while improved warning and evacuations should reduce loss of lives.

However, it is not clear that these costly plans make sense, as illustrated by the city of Kamaishi. The city, although already declining after its steel industry closed, was chosen for protection by a $1.6 billion breakwater, completed in 2008. A song produced by the government "Protecting Us for a Hundred Years" praised the structure: "It protects the steel town of Kamaishi, it protects our livelihoods, it protects the people's future." However, the breakwater collapsed when struck by the tsunami and 935 people in the city died, many of whom could have evacuated once warnings were given but did not, believing they were safe.

As the *New York Times* explained,

> After the tsunami and the nuclear meltdowns at Fukushima, some Japanese leaders vowed that the disasters would give birth to a new Japan, the way the end of World War II had done. A creative reconstruction of the northeast, where Japan would showcase its leadership in dealing with a rapidly aging and shrinking society, was supposed to lead the way. But as details of the government's reconstruction spending emerge, signs are growing that Japan has yet to move beyond a postwar model that enriched the country but ultimately left it stagnant for the past two decades. As the story of Kamaishi's breakwater suggests, the kind of cozy ties between government and industry that contributed to the Fukushima nuclear disaster are driving much of the reconstruction and the fight for a share of the $120 billion budget expected to be approved in a few weeks. The insistence on rebuilding breakwaters and sea walls reflects a recovery plan out of step with the times, critics say, a waste of money that aims to protect an area of rapidly declining population with technology that is a proven failure.

Instead, critics argue that it would be more efficient to relocate some communities inland, because their populations are small and decreasing. Otherwise, as explained in the *New York Times*, "in 30 years there might be nothing here but fancy breakwaters and empty houses." There is also new interest in exploring methods other than large-scale engineering works to enhance community resilience to tsunamis.

The situation in Japan illustrates how major policy choices are often made without careful analysis of their costs and benefits, and that it is often far from clear that they are wise. There is thus a growing sense among many interested in natural hazards that more rational methods for making policies are needed.

Questions

2.1. Use the data in Figure 2.1 and the results of the 2011 earthquake in section 2.1 to estimate the odds of being killed by an earthquake in Japan. Because these estimates depend in part on how recently a disastrous earthquake has occurred, estimate the odds both with and without the 2011 earthquake.

2.2. Compare the results from question 2.1 to those for the US derived in question 1.4. What factors do you think are the main causes of the difference?

2.3. Section 2.2 discussed factors that cause earthquake hazard maps to fail – bad physics, bad assumptions, bad data, and bad luck. With the advantage of 20/20 hindsight, classify the problems now recognized with the Tohoku map into these four categories and explain your choices.

2.4. After a major earthquake and tsunami like that which struck Tohoku in 2011, an important question in assessing how mitigation measures worked is how much damage resulted from the earthquake shaking compared to how much was caused by the tsunami. To gain some insight, watch some videos of the Tohoku tsunami linked *http://www .geologyinmotion.com/2011/03/more-videos-of-tsunami-and-situation -in.html*.

How would you describe the relative damage? Suggest possible causes and ways to test your hypotheses.

2.5. Based on the executive summary of the report of the Fukushima nuclear accident investigation commission at *http://warp.da.ndl.go.jp/info :ndljp/pid/3856371/naiic.go.jp/en/index.html*

what do you consider the most important actions that could have avoided the disaster? Why were they not taken?

Further Reading and Sources

Onishi (2011a,b,c) and Normile (2012) describe the Tohoku tsunami defenses and the communities' attitudes.The Japanese hazard map and information about it come from the Earthquake Research Committee (2009, 2010). Chang (2011), Geller (2011), Kerr (2011), Kossobokov and Nekrasova (2012), Sagiya (2011), Stein and Okal (2011), Stein et al. (2011, 2012), and Yomogida et al. (2011) discuss the Tohoku issues and whether the Japanese map's failure reflected bad luck or a systemic problem.

Taleb (2007) popularized the "black swan" terminology. Wyss et al. (2012) argue that many more fatalities occur worldwide than would have been expected from hazard maps.

Nöggerath et al. (2011) and Onishi and Fackler (2011) describe the history of tsunami mitigation measures for the Fukishima plant, and Onishi (2011d) quotes the safety myth song. The National Diet of Japan Fukushima Nuclear Accident Independent Investigation Commission report is described by Tabuchi (2012) and available at *http://warp.da.ndl.go.jp/info:ndljp/pid/385 6371/naiic.go.jp/en*. Ando et al. (2011) describe the response to the tsunami warning.

The *Science* magazine quote is from Kerr (2011). Cyranoski (2012), Fackler (2012), Normile (2012), and Onishi (2011c) discuss plans to rebuild Tohoku tsunami defenses, and Ewing and Synolakis (2010) discuss alternatives to seawalls.

References

Ando, M., M. Ishida, Y. Hayashi, and C. Mizuki, Interviews with survivors of Tohoku earthquake provide insights into fatality rate, *Eos Trans. AGU*, *46*, 411, 2011.

Chang, K., Blindsided by ferocity unleashed by a fault, *New York Times*, March 21, 2011.

Cyranoski, D., Rebuilding Japan, *Nature*, *483*, 141–143, 2012.

Earthquake Research Committee, Long-term forecast of earthquakes from Sanriku-oki to Boso-oki, 2009.

Earthquake Research Committee, National seismic hazard maps for Japan, 2010.

Ewing, L., and C. Synolakis, Community resilience, *Proceedings of the International Conference on Coastal Engineering*, *32*, 1–13, 2010.

Fackler, M., In Japan, a rebuilt island serves as a cautionary tale, *New York Times*, January 9, 2012.

Geller, R. J., Shake-up time for Japanese seismology, *Nature*, *472*, 407–409, 2011.

Kerr, R. A., Seismic crystal ball proving mostly cloudy around the world, *Science*, *332*, 912–913, 2011.

Kossobokov, V. G., and A. K. Nekrasova, Global seismic hazard assessment program maps are erroneous, *Seismic Instrum.*, *48*, 162–170, 2012.

Normile, D., One year after the devastation, Tohoku designs its renewal, *Science*, *335*, 1164–1166, 2012.

Nöggerath, J., R. J. Geller, and V. K. Gusiakov, Fukushima: the myth of safety, the reality of geoscience, *Bull. At. Sci.*, *67*, 37–46, 2011.

Onishi, N., Seawalls offered little protection against tsunami's crushing waves, *New York Times*, March 13, 2011a.

Onishi, N., In Japan, seawall offered a false sense of security, *New York Times*, March 31, 2011b.

Onishi, N., Japan revives a sea barrier that failed to hold, *New York Times*, November 2, 2011c.

Onishi, N., "Safety Myth" left Japan ripe for nuclear crisis, *New York Times*, June 24, 2011d.

Onishi, N., and M. Fackler, Japanese officials ignored or concealed dangers, *New York Times*, May 16, 2011.

Sagiya, T., Integrate all available data, *Nature*, *473*, 146–147, 2011.

Stein, S., and E. A. Okal, The size of the 2011 Tohoku earthquake needn't have been a surprise, *Eos Trans. AGU*, *92*, 227–228, 2011.

Stein, S., R. J. Geller, and M. Liu, Bad assumptions or bad luck: why earthquake hazard maps need objective testing, *Seismol. Res. Lett.*, *82*, 623–626, 2011.

Stein, S., R. J. Geller, and M. Liu, Why earthquake hazard maps often fail and what to do about it, *Tectonophysics*, *562–563*, 623–626, 2012.

Tabuchi, H., Inquiry declares Fukushima crisis a man-made disaster, *New York Times*, July 5, 2012.

Taleb, N. N., *The Black Swan: The Impact of the Highly Improbable*, Random House, New York, 2007.

Wyss, M., A. Nekraskova, and V. Kossobokov, Errors in expected human losses due to incorrect seismic hazard estimates, *Nat. Hazard.*, *62*, 927–935, 2012.

Yomogida, K., K. Yoshizawa, J. Koyama, and M. Tsuzuki, Along-dip segmentation of the 2011 off the Pacific coast of Tohoku Earthquake and comparison with other megathrust earthquakes, *Earth Planets Space*, *63*, 697–701, 2011.

3

Nature Bats Last

Glendower: "I can call spirits from the vasty deep."
Hotspur: "Why, so can I, or so can any man; but will they come when
you do call for them?"

Shakespeare, Henry IV

3.1 Prediction Is Hard

When natural hazard forecasts fail badly, as for the Tohoku earthquake, the public is often surprised. The public image of science is people in white coats using complicated machines to measure something. They measure it, and then they know it.

Assessing natural hazards is very different. A natural hazard, like that of earthquakes, or the probability of an event like an earthquake, is not something scientists can measure or know precisely. We cannot talk usefully – except in general terms – about "the" earthquake hazard or "the" probability of an earthquake. We can make various models, combining what we know and our ideas about how the earth works, and come up with a wide range of values. Although these estimates come from big computer programs and are displayed in colorful brochures and maps, sometimes they prove pretty good, and sometimes they do not.

Often what seemed like a reasonable assumption does not work very well. In these cases, the hazard assessment may not do much good – and can do

Playing against Nature: Integrating Science and Economics to Mitigate Natural Hazards in an Uncertain World, First Edition. Seth Stein and Jerome Stein.
© 2014 John Wiley & Sons, Ltd. Published 2014 by John Wiley & Sons, Ltd.
Companion Website: www.wiley.com/go/stein/nature

harm – for the communities affected. This is embarrassing, but that is the way it is. We are trying to predict how complicated physical systems that we do not fully understand will behave in the future. Not surprisingly, we are often wrong. In this game, nature bats last and does not have to do what we think it should. As in the Shakespeare quotation, we often overestimate how well we can predict what nature will do.

This tendency is not surprising, given that humans generally overestimate their ability to predict the future of almost anything, including natural processes, political events, or economics. As physicist Niels Bohr famously said, *"Prediction is very difficult, especially about the future."*

Many authors have looked at this issue, and come to the same conclusion. Dan Gardner's book *Future Babble: Why expert predictions fail and why we believe them anyway* explains, "They're wrong a lot, these experts. History is littered with failed predictions. Whole books can be filled with them. Many have." In *The future of everything: The science of prediction,* David Orrell says:

> While scientists have had great success in squinting through microscopes at the smallest forms of life, or smashing atoms together in giant particle accelerators, or using telescopes to look forward in space and backwards in time at the formation of distant galaxies, their visions into the future have been blurred and murky. As a result, projections tend to go astray.

Once we recognize that predictions often do poorly, understanding why can help us avoid some of the problems. Nate Silver, in *The signal and the noise: Why so many predictions fail but some don't,* observes that

> the most calamitous failures of prediction usually have a lot in common. We focus on the signals that tell a story about the world as we would like it to be, not how it really is . . . We make approximations and assumptions about the world that are much cruder than we realize. We abhor uncertainty, even when it is an irreducible part of the problem we are trying to solve.

People's attempts to explain why their predictions failed have similar problems. Silver explains:

> When you make a prediction that goes so badly, you have a choice of how to explain it. One path is to blame external circumstances – what we might think of as bad luck. Sometimes this is a reasonable choice or even the correct one. When the National Weather Service says there is a 90% chance of clear skies, but it rains instead and spoils your golf outing, you can't really blame them. Decades of historical data shows that when the Weather Service says there is a 1 in ten chance of rain, it really does rain about 10% of the time. This

explanation becomes less credible when the forecaster does not have a history of successful predictions and when the magnitude of his error is larger. In these cases, it is more likely that the fault lies with the forecaster's model of the world and not with the world itself.

3.2   Forecasts, Predictions, and Warnings

In mitigating natural hazards, researchers try to infer something useful about what may happen in the future, and then use these inferences to help communities prepare. The approaches used can be grouped in three general categories:

Long-term forecasts seek to describe hazards on a long time scale, rather than when or where a specific event will occur. For example, earthquake hazard maps try to predict the amount of ground shaking to expect at a certain probability level over a certain time – tens, hundreds, or thousands of years. Similarly, hurricane hazard maps predict the number of hurricanes expected to strike an area (Figure 3.1), or the resulting flooding or wind speeds. Volcanic hazard maps show the expected hazard due to effects such as mud flows (lahars), lava flows, or pyroclastic (hot rock and gas) flows (Figure 3.2). Such forecasts are used in planning mitigation strategies.

Short-term predictions try to predict if and how an imminent event will occur in the near future – days to months depending on the specific hazard – so that emergency measures can be prepared and taken. For example, once a volcano shows signs of unrest, such as increased seismicity, ground deformation, or gas emissions, volcanologists seek to predict whether an eruption is forthcoming. Similarly, once a major hurricane has formed, meteorologists try to predict its future track.

Real time warnings are given once an event is ongoing and immediate measures can be taken. Once a large earthquake that is likely to generate a tsunami occurs, people in potentially dangerous areas can be warned to evacuate to higher ground. Earthquake early warning systems can give few seconds to tens of seconds warning of strong ground shaking, which can be enough to shut down systems such as train networks and thereby reduce damage. Warning systems are also used for floods, tornadoes, and other hazards.

Distinguishing between forecasts, predictions, and warnings is useful for our discussions, although the categories somewhat overlap and the terminology is not standardized between different fields of science. Hurricane forecasts, for example, are predictions in the sense that they predict the evolution of a storm that already exists, even if it has not yet reached hurricane strength (winds above 74 miles/hr or 119 km/hr).

■ 20 to 40 strikes ■ 40 to 60 strikes ■ over 60 strikes
Hurricane potential in the United States and Puerto Rico

Figure 3.1   The number of hurricanes expected to strike various areas of the U.S. during a 100-year period based on historical data, showing areas expected to receive 20 to 40, 40 to 60, or more than 60 strikes. (USGS, 2005.)

All of these approaches are challenging and work to varying degrees, depending on the specific hazard and how well we understand it. Sometimes they work, and the results are good for society. Sometimes they fail, causing problems. The failures come in two kinds. A *false negative* is an unpredicted or underpredicted hazard, which can cause loss of life and property. A *false positive* is an overpredicted hazard, which can waste resources spent on excessive mitigation and cause businesses to locate elsewhere. False positives can also cause the public to ignore warnings, as in Aesop's ancient Greek fable of the boy who cried "wolf!" (Figure 3.3).

The Tohoku earthquake illustrated both problems. The false negative was that the possibility of a magnitude 9 earthquake and corresponding tsunami had not been recognized. On the other hand, some coastal residents ignored the tsunami warning – which proved correct and actually underestimated the tsunami size – because of past false positives. Researchers who interviewed

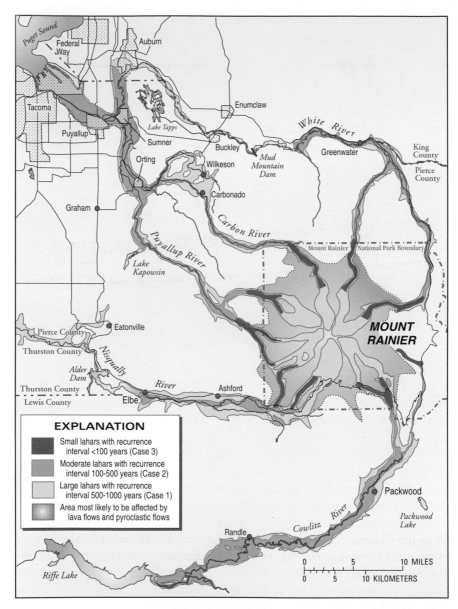

Figure 3.2   Volcanic hazard map for the Mount Rainier, Washington area. (Hoblitt et al., 1998.)

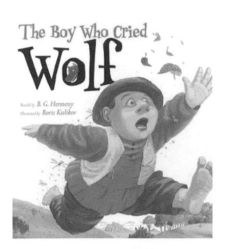
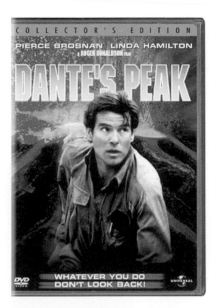

Figure 3.3   Issues of prediction and forecasting have interested the public from the 5th century BC to the present, as shown by the ancient tale of *The Boy Who Cried Wolf*, and the more modern take in the 1997 film *Dante's Peak* in which the volcano actually erupts.

residents noted that in the previous four years, sixteen warnings or alerts had issued for "small or even negligible tsunamis. These frequent warnings with overestimated tsunami height influenced the behavior of the residents." As a result, new techniques are being introduced to give better real-time estimates of tsunami heights.

Problems can also be organizational, as illustrated by an embarrassing false positive after the February 2010 Maule, Chile, earthquake. Although the tsunami generated by the magnitude 8.8 earthquake did considerable damage along the Chilean coast, tsunami forecasting that incorporated measurements of the wave from ocean buoys showed little danger to more distant communities. Nonetheless, the US government's Pacific Tsunami Warning Center issued warnings that led to the evacuation of coastal areas including downtown Honolulu. Residents picnicked high above the city in lawn chairs, watching for the great wave that didn't materialize. Fortunately, the economic losses were minor because it was Saturday. Still, as Costas Synolakis, one of the developers of the forecasting model, pointed out, "The authorities in charge need to listen to science," because "Every ounce of extra prevention is counterproductive as it reduces the overall credibility of the system."

Figure 3.4   1980 eruption of Mount St. Helens. (USGS/Cascades Volcano Observatory.)

Volcano prediction is sometimes very successful. The area around Mt. St. Helens, Washington, was evacuated before the giant eruption of May 1980 (Figure 3.4), reducing the loss of life to only 60 people, including a geologist studying the volcano and citizens who refused to leave. The largest eruption of the second half of the 20th century, that of Mt. Pinatubo in the Philippines in 1991, destroyed over 100,000 houses and a nearby US Air Force base, yet only 281 people died because of evacuations during the preceding days. The US Geological Survey (USGS) team's real-time decision making about predicting the eruption is vividly described in the film "*In the path of a killer volcano*."

In other cases a volcano seems to be preparing to erupt, but does not. In 1982, uplift of the volcanic dome and other activity near the resort town of Mammoth Lakes, California, suggested that an eruption might be imminent. A volcano alert was issued, causing significant problems. Housing prices fell

40%. In the next few years, dozens of businesses closed, new shopping centers stood empty, and townspeople left to seek jobs elsewhere. Angry residents called the US Geological Survey the "US Guessing Society," and the county supervisor who arranged for a new road providing an escape route in the event of an eruption was recalled in a special election. However, even in hindsight the alert seems sensible given the data then available, illustrating the challenge involved. The incident provided the basis for the film *Dante's Peak*, in which the volcano actually erupts (Figure 3.3).

When hazard assessments or mitigation measures fail, the authorities typically ignore, deny, excuse, or minimize the failure. Thus after the Tohoku earthquake, many claims were made that such an event could not have been foreseen. Although this is a natural human response, it is more useful to analyze failures to understand what went wrong and improve the assessment and mitigation process. This situation is like what happens after an airplane crash – although it is tempting to blame bad luck, it is more useful to focus on the problems that caused the crash and how to correct them.

3.3   Earthquake Prediction

How well forecasts, predictions, or warnings work depends on how well the underlying processes are understood, how easy it is to observe them, and the time scales involved. Comparing some different examples shows a pattern one might expect (Figure 3.5).

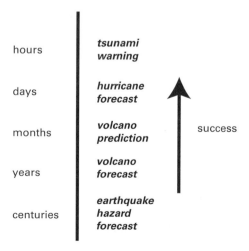

Figure 3.5   Schematic comparison of some forecast and warning methods.

Tsunami warnings generally do well because the model predicts the behavior of a process – tsunami generation and propagation – whose science is something fairly well understood on a time scale of hours after an earthquake has happened. Hurricane or other weather forecasts do fairly well, predicting how the atmosphere will evolve over a few days. Predicting how a volcanic system of magma and gas underground will evolve over days to months is harder, because it is difficult to observe, but still often succeeds. Results for these and some other natural hazard applications are getting steadily better due to new observational technologies and modeling methods. For example, a more sophisticated computer model operated by a European consortium did better than US models at predicting the evolution of the complicated Hurricane Sandy that struck the New York area in 2012.

In contrast, progress for earthquakes has been much slower. As we have seen, successfully forecasting earthquake hazards over tens to hundreds of years is very difficult. Moreover, one of the most tantalizingtools for natural hazard mitigation, earthquake prediction, has been unsuccessful despite decades of expensive efforts. "Prediction" would involve identifying where and when earthquakes will strike within a few days or weeks, compared to "forecasting" for the much longer time scales involved in hazard maps. Prediction is even harder than long-term forecasting. A common analogy is that although a bending stick will eventually snap, it is hard to predict exactly when. To do so requires either a theoretical basis for knowing when the stick will break, given a history of the applied force, or observing some change in physical properties that immediately precedes the stick's failure.

In the 1960s and 1970s, well-funded government earthquake prediction programs began in the US, China, Japan, and the USSR. These programs relied on two approaches. One was based on laboratory experiments showing changes in the physical properties of rocks prior to fracture, implying that earthquake precursors could be identified. A second was the idea of the seismic cycle, in which strain accumulates over time following a large earthquake (Figure 2.2). Hence areas on major faults that had not had recent earthquakes could be considered "seismic gaps" likely to have large earthquakes.

The idea that earthquake prediction was about to become reality was promoted heavily in the media. US Geological Survey director William Pecora announced in 1969, "We are predicting another massive earthquake certainly within the next 30 years and most likely in the next decade or so," on the San Andreas fault. Louis Pakiser of the USGS announced that if funding were granted, scientists would "be able to predict earthquakes in

Figure 3.6　The U.S. National Earthquake Hazards Reduction Program (NEHRP) was established in 1977 in response to claims like that in 1976 by the National Research Council, the operating arm of the National Academy of Sciences, that "it is the panel's unanimous opinion that the development of an effective earthquake-prediction capability is an achievable goal."

five years." Seismologist Lynn Sykes stated that "prediction of some classes of earthquakes in some tectonic provinces is virtually a reality." California senator Alan Cranston, prediction's leading political supporter, told reporters that "we have the technology to develop a reliable prediction system already at hand." Although Presidential science advisor Donald Hornig questioned the need for an expensive program given the low death rate from earthquakes, lobbying prevailed. Funding poured into the US National Earthquake Hazards Reduction Program (Figure 3.6) and similar programs in other countries.

The US efforts produced two spectacular failures. In 1975, while the bill for earthquake prediction funding was stalled in Congress, USGS researchers reported 30–45 cm of uplift along the San Andreas fault near Palmdale, California. This highly publicized "Palmdale Bulge" was interpreted as evidence for an impending large earthquake. USGS director Vincent McKelvey expressed his view that "a great earthquake" would occur "in the area presently affected by the . . . 'Palmdale Bulge' . . . possibly within the next decade" that might cause up to 12,000 deaths, 48,000 serious injuries, 40,000 damaged buildings, and up to $25 billion in damage. The California Seismic Safety Commission stated that "the uplift should be considered a possible threat to public safety" and urged immediate action to prepare for a possible disaster. News media joined the cry. However, the earthquake did not occur, and reanalysis of the data implied that the bulge had been an artifact of errors involved in referring the vertical motions to sea level via a traverse across the San Gabriel mountains. In Hough's (2009) words, the Bulge changed to "the Palmdale soufflé – flattened almost entirely by careful analysis of data."

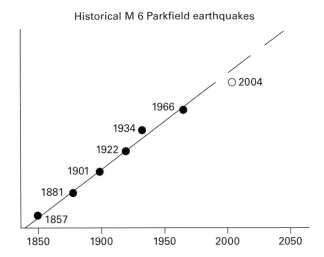

Figure 3.7   The Parkfield earthquake predicted to occur within five years of 1988 occurred in 2004. Black dots show the dates of earthquakes before the prediction was made, and the line shows when earthquakes 22 years apart should happen. (Stein, 2010. Reproduced with permission of Columbia University Press.)

The program culminated in 1985 when the USGS launched an official earthquake prediction experiment near Parkfield, California, a town of about 20 people whose motto is "Be here when it happens." This part of the San Andreas fault had had magnitude 6 earthquakes about every 22 years, with the last in 1966 (Figure 3.7). Thus the USGS predicted at the 95% confidence level that the next such earthquake would occur within five years of 1988, or before 1993. Seismometers, strainmeters, creepmeters, GPS receivers, tiltmeters, water level gauges, electromagnetic sensors, and video cameras were set up to monitor what would happen before and during the earthquake. The USGS's National Earthquake Prediction Evaluation Council, a group of university scientists who are funded by and work closely with the USGS, endorsed the highly publicized $20 million "Porkfield" project. The *Economist* magazine commented, "Parkfield is geophysics' Waterloo. If the earthquake comes without warnings of any kind, earthquakes are unpredictable and science is defeated. There will be no excuses left, for never has an ambush been more carefully laid."

Exactly that happened. The earthquake did not occur by 1993, leading *Science* magazine to conclude, "Seismologists' first official earthquake forecast has failed, ushering in an era of heightened uncertainty and more modest ambitions." A likely explanation was that the uncertainty in the repeat time had been underestimated by discounting the fact that the 1934 earthquake did not fit the pattern well (Figure 3.7). Although a USGS review committee criticized "the misconception that the experiment has now somehow failed," the handwriting was on the wall.

An earthquake eventually occurred near Parkfield in 2004, eleven years after the end of the prediction window, with no detectable precursors that could have led to a short-term prediction. It is unclear whether the 2004 event should be regarded as the predicted earthquake coming too late, or just the next earthquake on that part of the fault. For that matter, we do not know whether the fact that earthquakes occurred about 22 years apart for a while reflects an important aspect of the physics of this particular part of the San Andreas, or just an apparent pattern that arose by chance given that we have a short history and many segments of the San Andreas are of similar length. After all, flipping a coin enough times will give some impressive-looking patterns of heads or tails. With only a short set of data, we could easily interpret significance to what was actually a random fluctuation.

Prediction attempts elsewhere have had the same problem. Although various possible precursors have been suggested, no reliable and reproducible ones have been found. Some may have been real in certain cases, but none have yet proved to be a general feature preceding all earthquakes, or to stand out convincingly from the normal range of the earth's variable behavior. It is tempting to note a precursory pattern after an earthquake based on a small set of data, and suggest that the earthquake might have been predicted. However, rigorous tests with large sets of data are needed to tell whether a possible precursory behavior is real and correlates with earthquakes more frequently than expected purely by chance. Most crucially, any such pattern needs to be tested by predicting future earthquakes. For example, despite China's major national prediction program, no anomalous behavior was identified before the 2008 Wenchuan earthquake. Similarly, the seismic gap hypothesis has not yet proven successful in identifying future earthquake locations significantly better than random guessing.

As a result, seismologists have largely abandoned efforts to predict earthquakes on time scales of less than a few years, although the ongoing government program in Japan to try to issue a prediction within three days of an anticipated magnitude 8 earthquake in the Tokai region (Figure 2.1) is an exception.

The frustrating attempts over the years to predict earthquakes have led seismologists to famous aphorisms: Charles Richter's that "only fools and charlatans predict earthquakes" and Hiroo Kanamori's that "it is hard to predict earthquakes, especially before they happen." For our purposes, these failures remind us that "nature bats last" – it is not obligated to act the way

people, even in well-funded and enthusiastically promoted government programs, expect.

3.4   Chaos

The failure of earthquake prediction efforts gives insight into how earthquakes
work. The most popular hypothesis is that all earthquakes start off as tiny
earthquakes, which happen frequently, but only a few cascade via a random
failure process into successively larger earthquakes. This hypothesis draws
on ideas from nonlinear dynamics or chaos theory, in which some small
perturbations grow to have unpredictable large consequences.

These ideas were put forward in 1963 by meteorologist Edward Lorenz,
who was surprised to find that small changes in simple computer models of
the weather could give very different results. Assuming that the real atmosphere acts this way, small perturbations could grow to have large effects.
Lorenz described this effect using the famous analogy that the flap of a but-
terfly's wings in Brazil might set off a tornado in Texas. This concept reached
the public in the movie "Jurassic Park," where small problems grew into big
ones that made the dinosaur park collapse.

A simple illustration of this idea comes from considering a system whose
evolution in time is described by the equation $x(t + 1) = ax(t)^2 - 1$ (Figure
3.8). For $a = 2$, two runs starting off at time $t = 0$ with slightly different
values, $x(0) = 0.750$ and $x(0) = 0.749$, yield curves that differ significantly
within a short time. Different values of a give other complicated behavior.

The fact that small differences grow is part of the reason why weather
forecasts get less accurate as they project further into the future – tomorrow's
forecast is much better than one for the next five days. By about two weeks,
the uncertainties are so large that forecasts are not useful.

An interesting thought experiment, suggested by Lorenz (1995), is to ask
what the weather would be like if it were not chaotic. In this case, weather
would be controlled only by the seasons, so year after year storms would
follow the same tracks, making planning to avoid storm damage easy. In
reality, storms are very different from year to year (Figure 3.9). Thus, in
Lorenz's words, "the greater difficulty in planning things in the real world,
and the occasional disastrous effects of hurricanes and other storms, must
therefore be attributed to chaos."

If earthquakes behave analogously, there is nothing special about those tiny
earthquakes that happen to grow into large ones. As a result, no observable
precursors should occur before the large earthquakes. Because the earth does
not know which tiny earthquakes will become big ones, it cannot tell us. If

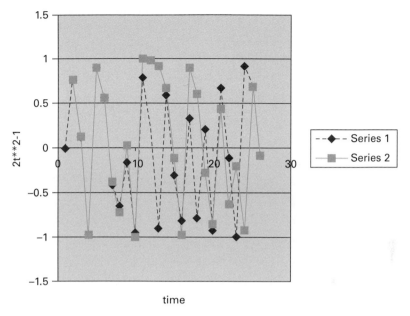

Figure 3.8 Small perturbations grow. Comparison of two time series generated by the same equation, showing how slightly different initial conditions quickly lead to quite different values.

so, earthquake prediction is either impossible or nearly so. Support for this view comes from the failure to observe a compelling pattern of precursors before earthquakes.

By analogy to Lorenz's hurricane thought experiment, without chaos the steady motion between plates should produce a pattern of earthquakes that repeat in space and time. In contrast, the chaos view predicts that the locations of big earthquakes on a plate boundary and interval between them should be highly variable. As Figure 2.4 demonstrates for the Nankai Trough, this is what the geological record shows. This variability is why using seismic gaps to predict earthquake locations often fails.

The chaos view does not contradict the idea that big earthquakes happen when strain is released after being built up on locked faults over the years by plate motion. In Lorenz's view, the overall frequency of storms depends on the energetics of the atmosphere, but minuscule disturbances can modify when they occur. Similarly, this view implies that over many years the total slip in earthquakes on a plate boundary like the Japan Trench is determined by the motion between the plates, but the history of where and when large earthquakes occur and how big they are can be very complicated.

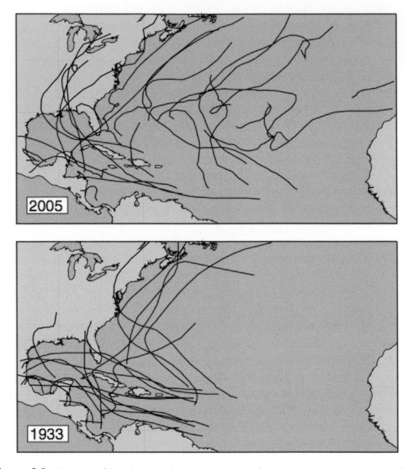

Figure 3.9   Tracks of North Atlantic hurricanes, tropical storms, and depressions for two very active hurricane seasons. (Ebeling and Stein, 2011. Reproduced with permission of the Seismological Society of America.)

This argument brings up the issue of what it means to say that something is "random." We usually think of randomness in terms of an operation like throwing a die, where there is an equal probability of getting each number between one and six. If we throw the die many times, each number will come up equally often. In mathematical terms, what number comes up is described by a random variable. However, throwing a die is really a physical process. The value we get depends on how we throw the die, the properties of the die, and the properties of the surface it lands on. In principle, we could describe these factors and use Newtonian mechanics to predict what value will come up. In practice, what happens is so complicated to describe that we treat the outcome as random.

We think of the complicated systems generating natural hazards – earthquakes, volcanoes, storms, etc. – as deterministic in the sense that if we understood their physics well enough and knew the initial conditions at some time, we ideally could predict their future. However, as the example in Figure 3.8 shows, some very simple deterministic systems are essentially impossible to predict in the sense that solutions starting off close to each other evolve very differently. In such cases, because we cannot make adequate deterministic models, we sometimes make models including random variables, called *stochastic models*. In these models, saying something is "random" means "so complicated that we don't understand the physics, so instead of trying to predict it deterministically based on a model of the physics, we treat it as described by a random variable." Stochastic models are used in many applications, such as describing how often an earthquake of a given size will happen.

It is important to remember that these stochastic models are attempts to describe the behavior of systems that we do not understand well enough to model deterministically. The parameters of stochastic models are derived by fitting the model to data that we have, such as an earthquake history. However, because the real system is more complicated than the model, often the model fails to predict future behavior. Although we can then make an even more complicated model including more terms, we are not guaranteed that the new model will describe the real world better, so its predictions may still fail. The series of updated earthquake hazard maps for Italy (Figure 1.6) show this problem.

Questions

3.1. If you were allocating your nation's earth science research budget, would you use some of the funds for research on predicting earthquakes? Why or why not?

3.2. In hindsight, when home prices fell sharply in Mammoth Lakes after the volcano alert, it was a great opportunity to invest. However, hindsight is easy. At the time, would you have bought a vacation home there? Why or why not?

3.3. Using an Excel spreadsheet or computer program, compute the intervals between the dates of the Parkfield earthquakes given in Figure 3.7. Next, find the mean and standard deviation of these intervals. With Excel, these can be done using the AVERAGE and STDEV functions. First, use only the events before the 1988 prediction. Second, use those events but assume that the 1934 earthquake had occurred in 1944, which is essentially what was done when the prediction was made. Third, use the actual dates including the 2004 earthquake. Compare the results.

3.4. How might you try to determine whether the fact that earthquakes near Parkfield occurred for a while about 22 years apart reflects an important aspect of the physics of this particular part of the San Andreas, or just represents an apparent pattern that arose by chance?

3.5. Use an Excel spreadsheet or computer program to compute the time series evolution shown in Figure 3.8. Compare starting values of 0.7500 and 0.7490, and values of 0.7500 and 0.7499. What are the differences?

3.6. Flip a coin twenty times and record the results. Then, take this series, call heads "1" and tails "0," and plot the series that results. Next, imagine that you did not know where the data came from, and note any interesting patterns you see. What might you conclude if you only had part of the series? Finally, imagine fitting a complicated function to the series. How often would you expect this model to correctly predict the next flip?

Further Reading and Sources

Earthquake early warning systems are described at *http://www.elarms.org*. An overview of tsunami warning systems is at *http://www.tsunami.noaa.gov/ warning_system_works.html*.

Discussion of the responses to the Tohoku and Chile tsunami warnings are from Schiermeier (2010) and Ando et al. (2011). Francis and Oppenheimer (2004), Marzocchi and Woo (2009) and *http://volcanoes.usgs.gov/* discuss volcano forecasting.

The trailer for the film "Dante's Peak" is at *http://www.reelz.com/trailer -clips/22273/dantes-peak-trailer*.

Blakeslee (1990) and Monastersky (1991) describe the Mammoth Lakes prediction. Figure 3.4 (photograph of the 1980 eruption of Mount St. Helens) is from *www.fs.fed.us/pnw/mtsthelens/photo-gallery*. Emanuel (2012) discusses forecasts of Hurricane Sandy. Geller (1997), Geschwind (2001), and Hough (2009) describe the history of earthquake prediction.

Kerr (1984, 1985, 1986, 1992, 1993, 2004) and Finn (1992) describe the Parkfield prediction's rise and fall, and a video about it is posted at *http:// www.youtube.com/watch?v=Av1V_6un_Oo*. The *Economist* quotation is from August 1, 1987. The report defending the unsuccessful prediction is Hager et al. (1994). Savage (1993) discusses the prediction's statistical difficulties, and

Bakun et al. (2005) and Jackson and Kagan (2006) discuss the 2004 earthquake.

The Richter quotation is from Hough's (2007) biography. Chen and Wang (2010) discuss the Chinese prediction program, Geller (2011) discusses the Japanese program, and Kagan and Jackson (1991, 1995) discuss the gap hypothesis.

Lorenz (1995) gives a nice introduction to chaos. The "Jurassic Park" chaos scene is posted at *http://movieclips.com/fiwF-jurassic-park-movie -chaos-theory/*.

References

Ando, M., M. Ishida, Y. Hayashi, and C. Mizuki, Interviews with survivors of Tohoku earthquake provide insights into fatality rate, *Eos Trans. AGU*, *46*, 411, 2011.

Bakun, W. H., et al., Implications for prediction and hazard assessment from the 2004 Parkfield earthquake, *Nature*, *437*, 969–974, 2005.

Blakeslee, S., Wrong once, experts keep quiet on volcanic activity in California, *New York Times*, September 11, 1990.

Chen, Q.-F., and K. Wang, The 2008 Wenchuan earthquake and earthquake prediction in China, *Bull. Seismol. Soc. Am.*, *100*, 2840–2857, 2010.

Ebeling, C., and S. Stein, Seismological identification and characterization of a large hurricane, *Bull. Seismol. Soc. Am.*, *101*, 399–403, 2011.

Emanuel, K., Why America has fallen behind the world in storm forecasting, *Wall Street Journal*, October 28, 2012.

Finn, R., Rumblings grow about Parkfield in wake of earthquake prediction, *Nature*, *359*, 761, 1992.

Francis, P., and C. Oppenheimer, *Volcanoes*, Oxford Univ. Press, Oxford, 2004.

Gardner, D., *Future Babble: Why Expert Predictions Fail and Why We Believe Them Anyway*, McClelland & Stewart, Toronto, 2010.

Geller, R. J., Earthquake prediction: a critical review, *Geophys. J. Int.*, *131*, 425–450, 1997.

Geller, R. J., Shake-up time for Japanese seismology, *Nature*, *472*, 407–409, 2011.

Geschwind, C.-H., *California Earthquakes: Science, Risk, and the Politics of Hazard Mitigation*, Johns Hopkins University Press, Baltimore, MD, 2001.

Hager, B. H., C. A. Cornell, W. M. Medigovich, K. Mogi, R. M. Smith, L. T. Tobin, J. Stock, and R. Weldon, *Earthquake Research at Parkfield, California, for 1993 and Beyond*, Report of the NEPEC [National Earthquake Prediction Evaluation Council] Working Group, U.S. Geological Survey Circular, 1116, 1994.

Hoblitt, R., J. Walder, C. Driedger, K. Scott, P. Pringle, and J. Vallance, *Volcano Hazards from Mt Rainer, Washington*, U.S. Geological Survey Open-File Report 98-428, 1998.

Hough, S., *Predicting the Unpredictable*, Princeton University Press, Princeton, NJ, 2009.

Hough, S. E., *Richter's Scale: Measure of an Earthquake, Measure of a Man*, Princeton University Press, Princeton, NJ, 2007.

Jackson, D. D., and Y. Y. Kagan, The 2004 Parkfield earthquake, the 1985 prediction, and characteristic earthquakes: lessons for the future, *Bull. Seismol. Soc. Am.*, *96*, S397–S409, 2006.

Kagan, Y. Y., and D. D. Jackson, Seismic gap hypothesis: ten years after, *J. Geophys. Res.*, *96*(21), 419–421, 431, 1991.

Kagan, Y. Y., and D. D. Jackson, New seismic gap hypothesis: five years after, *J. Geophys. Res.*, *99*, 3943–3959, 1995.

Kerr, R., Seismologists issue a no-win earthquake warning, *Science*, *258*, 742–743, 1992.

Kerr, R. A., Stalking the next Parkfield earthquake, *Science*, *223*, 36–38, 1984.

Kerr, R. A., Earthquake forecast endorsed, *Science*, *228*, 311, 1985.

Kerr, R. A., Parkfield earthquake looks to be on schedule, *Science*, *231*, 116, 1986.

Kerr, R. A., Parkfield quakes skip a beat, *Science*, *259*, 1120–1122, 1993.

Kerr, R. A., Parkfield keeps secrets after a long-awaited quake, *Science*, *306*, 206–207, 2004.

Lorenz, E., *The Essence of Chaos*, University of Washington Press, Seattle, WA, 1995.

Marzocchi, W., and G. Woo, Principles of volcanic risk metrics: theory and the case study of Mount Vesuvius and Campi Flegrei, Italy B03213, *J. Geophys. Res.*, *114*, doi: 10.1029/2008JB005908, 2009.

Monastersky, R., Perils of prediction: are scientists prepared to warn the public about geologic hazards? *Sci. News*, *139*, 376, 1991.

Orrell, D., *The Future of Everything: The Science of Prediction*, Thunder's Mouth, New York, 2007.

Savage, J. C., The Parkfield prediction fallacy, *Bull. Seismol. Soc. Am.*, *83*, 1–6, 1993.

Schiermeier, Q., Model response to Chile quake, *Nature*, *464*, 14–15, 2010.

Silver, N., *The Signal and the Noise*, Penguin, New York, 2012.

Stein, S., *Disaster Deferred: How New Science is Changing our View of Earthquake Hazards in the Midwest*, Columbia University Press, New York, 2010.

4

Uncertainty and Probability

"I can live with doubt and uncertainty. I think it's much more interesting to live not knowing than to have answers which might be wrong."

Richard Feynman[1]

4.1 Basic Ideas

The examples in the previous chapters show that our biggest challenge in mitigating natural hazards comes from their uncertainty. We do not know where and when a major hazardous event will happen and what will result when it does. Thus instead of preparing for a known future, we have to prepare for an uncertain one. This means using our best guesses based on a combination of what we know about how the earth works and what we think about how the earth works. However, there's a lot that we don't know and some of what we think we know turns out to be wrong. As a result, sometimes our guesses work out well – and sometimes they don't.

Emotionally, this is unsatisfying, because people like certainty. However, we have to deal with the world as it is, not as we'd like it to be. To do this, we need to think in terms of *probability*, a branch of mathematics that deals with uncertainty. Probability lets us define terms such as "likely" or "probable" and work with them. Because probability is so crucial for our discussions, we introduce in this chapter some basic ideas that we will use. The suggested reading gives some sources for more information.

[1] Richard Feynman, *Horizon* interview, BBC, 1981.

Playing against Nature: Integrating Science and Economics to Mitigate Natural Hazards in an Uncertain World, First Edition. Seth Stein and Jerome Stein.
© 2014 John Wiley & Sons, Ltd. Published 2014 by John Wiley & Sons, Ltd.
Companion Website: www.wiley.com/go/stein/nature

Probability differs from the usual way mathematics is used in science and engineering. Traditionally, science uses mathematics in equations like Newton's laws of motion, which tell us what will happen. There are exact and right answers. In contrast, probability considers how likely different outcomes are.

The roots of the theory of probability are in games of chance. In 1654, a French gambler, the Chevalier de Mere, turned to his friend Blaise Pascal to figure out why he kept losing money in dice games. To solve his problem, Pascal invented a new branch of mathematics. Hence probability is easiest to explain in terms of gambling, although it is now used in any situation involving uncertainty. Natural hazards are an obvious application, because society is playing a high-stakes game of chance against nature, and trying to come up with sensible strategies to minimize its losses.

For example, we are fairly confident that San Francisco will eventually experience a major earthquake like that of 1906, that New Orleans will be struck again by a hurricane like Katrina or even larger, and that Vesuvius will someday erupt as it did in Roman times. What is harder is to say when these events may happen, which is crucial for hazard mitigation. Knowing that a major event may happen sometime in the next thousand years is much less useful than knowing it is likely in the next hundred. Measures that would be hard to justify in the first case might make sense in the second.

Ideally, we would like to understand these processes well enough to say in advance when an event will happen. However, we do not know how to do this, and in many cases it is doubtful that it is possible. As we discussed in section 3.4, in many cases nature seems to be chaotic. If so, we may never be able to predict with reasonable specificity when a big earthquake will happen or when a hurricane will form. The best we can do is make long-term forecasts by estimating the probability of a major event.

We do this using the idea of a random process, whose outcome cannot be predicted exactly, but only described in statistical terms. The event we are interested in is described by a *random variable*, a variable whose value depends on the random process. The variable could be binary – 1 if an earthquake, flood, hurricane, etc. happens and 0 if it does not. It could be an integer, such as the number of floods of a certain size in a certain time. It could also be continuous, such as the time between major earthquakes on a fault.

To formalize this concept, consider rolling a die N times. Each possible outcome is called an *event*. For example, we can call getting a six event A. After N rolls of which a number n have come up six, we define the frequency of event A as the fraction of the rolls that gave A:

$$f = n/N. \tag{4.1}$$

Note that f ranges between 0 (if no sixes came up) and 1 (if all the rolls gave sixes).

If we do this a number of times, the fraction f will vary because the outcome of each toss is random. However, as N, the number of rolls increases, f will approach a value called the *probability of A*:

$$P(A) = n/N. \tag{4.2}$$

The principle that as the number of rolls, or trials, gets large the fraction will converge to a specific value is called the *law of large numbers*. If the die is fair, so all six outcomes are equally likely, f will converge to 1/6 and $P(A) = 1/6$.

The probability that A will not happen, written $P(\bar{A})$, is defined by the fraction of trials that produced another outcome:

$$P(\bar{A}) = (N - n)/N = 1 - n/N = 1 - P(A). \tag{4.3}$$

Because we are certain ($P = 1$) that something will either happen or it won't,

$$P(A) + P(\bar{A}) = 1. \tag{4.4}$$

In this case, $P(A) = 1/6$ and $P(\bar{A}) = 5/6$, because five of the six equally likely outcomes are values other than six.

For natural hazards, we want to prepare for a range of possible events, each with a different probability and different consequences. Typically the larger events – floods, hurricanes, earthquakes, etc. – are rarer but more damaging. Thus we often consider the *expected value*, or *expectation*, of a random variable, which tells us what to expect if we wait long enough.

To define this, assume that a random variable x can have n outcomes x_1, x_2, x_3 ... x_n that have probabilities p_1, p_2, p_3 ... p_n. The expected value of x is defined as

$$E(x) = p_1 x_1 + p_2 x_2 + p_3 x_3 + \ldots + p_n x_n = \sum_{i=1}^{n} p_i x_i \tag{4.5}$$

which is the sum of the outcomes weighted by their probabilities. For example, a roulette wheel has 18 red numbers, 18 black numbers, and two green ones, 0 and 00. If we bet \$1 on red and the ball lands on red, we win another dollar. Otherwise, we lose the dollar. If x is a random variable describing the results, its expected value is

$$E(x) = (18/38) \times (1) + (20/38) \times (-1) = -0.053. \tag{4.6}$$

The negative expected value means that on average, over time, we lose 5 cents for every dollar we play. We will win some of the time, but are sure to lose in the long run. Hence the rational strategy is not to play, or to only play for fun while expecting to lose. The 5% profit goes to the casino.

We use similar expected value calculations in our game against nature. For example, assume that we anticipate that an area will be struck by magnitude 5, 6, and 7 earthquakes about once every 10, 100, and 1,000 years, and that these would cause about $10 million, $200 million, and $5 billion in damage. The annual probabilities of these earthquakes would be 1/10, 1/100, and 1/1000, so the expected damage per year would be the sum of the expected damage in each event times its probability:

$$E(x) = \frac{\$10m}{10} + \frac{\$200m}{100} + \frac{\$5000m}{1000} = \$8m. \tag{4.7}$$

This would be a long-term average, in that most years we expect no damage, in some years we expect modest damage, and we expect major damage in rare large earthquakes. As discussed later, this approach helps us to decide how much to invest in mitigation, which would reduce the losses.

4.2   Compound Events

We are often interested in the occurrence of two events. For example, what is the probability that a hurricane will strike an area two years in a row? To illustrate the approach used, consider the probability that both a die rolled will give a 6 and a coin tossed will come up heads. To describe these events, we call the die coming up six event A, and the coin coming up heads event B. Because the die can come up six ways and there are two ways for the coin to land, there are $2 \times 6 = 12$ possible outcomes, which are known as the *sample space*.

If we roll the die and toss the coin N times, the outcomes can be grouped into four categories:

n_1 the number in which A occurs but not B
n_2 the number in which B occurs but not A
n_3 the number in which both A and B occur
n_4 the number in which neither A nor B occur

If N is large, good estimates of the probabilities of the two individual events are

$$P(A) = (n_1 + n_3)/N, \qquad (4.8)$$

the probability of the die coming up six, and

$$P(B) = (n_2 + n_3)/N, \qquad (4.9)$$

the probability of a head.

Combining these gives the probability of compound events. The probability of both A and B happening, known as the *joint probability* of A and B, is

$$P(AB) = n_3/N \qquad (4.10)$$

and the probability of *either A, B, or both* happening is

$$P(A \cup B) = (n_1 + n_2 + n_3)/N. \qquad (4.11)$$

Comparing these shows that

$$P(A \cup B) = P(A) + P(B) - P(AB). \qquad (4.12)$$

This says that the probability that either A or B will happen is the sum of their probabilities minus the chance that both will happen. The subtraction is needed because otherwise we would have double counted.

Events A and B could be related in several ways. One is that they are *mutually independent*, so both cannot happen. For example, either a town will be struck by a hurricane next year or it won't. In this case, $P(AB) = 0$ and (4.12) becomes

$$P(A \cup B) = P(A) + P(B), \qquad (4.13)$$

so the probability that either A or B will happen is the sum of their probabilities.

Sometimes one event happening makes another more or less likely. For example, the probability of landslides is higher after heavy rains. This situation is described by the *conditional probability* $P(A|B)$ that event A will occur, given that event B has occurred. *Bayes' theorem* says that this conditional probability is

$$P(A|B) = P(AB)/P(B). \qquad (4.14)$$

This says that the conditional probability $P(A|B)$ equals the joint probability $P(AB)$, the probability that both A and B occur, divided by $P(B)$, the probability of B. The fact that B has occurred makes the conditional probability $P(A|B)$ higher than the joint probability $P(AB)$, because $P(B)$ is less than one.

Some events are *independent*, in that one occurring does not change the probability of the other. For example, the probability that a hurricane will strike an area in a year is not affected by whether a large earthquake strikes the area during that time. In such cases, the conditional probability of A given B, $P(A|B)$, is the same as the probability of A, $P(A)$, because event B has no effect. Bayes' theorem becomes

$$P(A|B) = P(AB)/P(B) = P(A). \qquad (4.15)$$

Rearranging terms shows the important result

$$P(AB) = P(A)P(B), \qquad (4.16)$$

which says that the joint probability of two independent events is the product of their probabilities. Similarly, for compound events composed of many events, the joint probability is the product of all their probabilities. For example, if A, B, C, and D are independent events,

$$P(ABCD) = P(A)P(B)P(C)P(D). \qquad (4.17)$$

This equation gives an easy way of finding out the probability that something will happen at least once, because it is 1 minus the probability that it will never happen.

$$P(at\ least\ once) = 1 - P(it\ will\ never\ happen). \qquad (4.18)$$

For example, the probability of getting at least one head in two coin tosses is 1 minus the probability that both tosses come up tails. Because the tosses are independent, that probability is the product of the probabilities that each toss comes up tails, so

$$P(a\ head) = 1 - P(all\ tails) = 1 - \left(\frac{1}{2} \times \frac{1}{2}\right) = \frac{3}{4}. \qquad (4.19)$$

Although you might think that because there is a 50% chance of getting a head on each toss, there is bound to be one in two tosses, that is not so. In fact, in three tosses the chance of a head is only

$$P(a\ head) = 1 - P(all\ tails) = 1 - \left(\frac{1}{2} \times \frac{1}{2} \times \frac{1}{2}\right) = 7/8 = 0.875 \qquad (4.20)$$

and in four tosses it is

$$P(a\ head) = 1 - P(all\ tails) = 1 - \left(\frac{1}{2} \times \frac{1}{2} \times \frac{1}{2} \times \frac{1}{2}\right) = 15/16 = 0.938. \qquad (4.21)$$

This concept is very important for natural hazards. A classic case involves the "100-year" flood, which is the height of flooding expected on average once every 100 years. The probability that it will happen in any one year is $1/100 = 0.01$, so the probability that it will not is 0.99. Thus in a hundred years the chance of one such flood is one minus the chance of none:

$$P(a\ flood) = 1 - P(no\ floods) = 1 - (0.99)^{100} = 1 - 0.37 = 0.63. \qquad (4.22)$$

So in 100 years there is a 63%, not 100%, chance of the 100-year flood. Equivalently, there is a 37% chance that the largest flood in the past 100 years is not the 100-year flood. We will explain this result further in Chapter 8.

This analysis assumes that the events are independent. As we will see in Chapter 6, many disasters occur because events that were assumed to be independent proved not to be. As a result, the very low probability estimated by multiplying the probabilities underestimated the real hazard.

To illustrate these ideas, imagine drawing cards from a deck. On the first draw, the probability of drawing a heart, which we call event B, is $13/52 = 1/4 = 0.25$. If the card is replaced, the probability of drawing a second heart, which we call event A, is also 0.25. In this case, because A and B are independent, the probability of drawing two successive hearts is

$$P(AB) = P(A)P(B) = 1/4 \times 1/4 = 1/16 = 0.063. \qquad (4.23)$$

However, if the first card drawn is not replaced, the draws are not independent, because the results of the first draw affect the second. The conditional probability of drawing a second heart given that one was drawn already is

$$P(A|B) = 12/51 = 0.235, \qquad (4.24)$$

which is less than 0.25, because the deck now has one less heart. Because events A and B are not independent, the probability of drawing two hearts is

$$P(AB) = P(A|B)P(B) = 0.235 \times 0.25 = 0.059, \qquad (4.25)$$

which is less than the probability (equation 4.23) if the draws were independent.

The strategies by which gamblers can win at blackjack (section 1.1) rely on keeping track of how the conditional probability of drawing valuable cards changes as successive cards are drawn. The game's goal is to get cards whose totals are closer to 21 than the dealer's, but not over 21. Cards from 2 through 9 are worth their face value, the jack, queen, and king are worth 10, and the ace is 1 or 11. On average high-value cards benefit the player, whereas low-value ones favor the dealer. Card-counting players keep track of the cards played in previous rounds, and so know whether the chances of drawing a valuable card are higher or lower. The dealer can respond by shuffling the deck after each round, so successive rounds are independent.

An analogous situation occurs for natural hazards, in which nature is the dealer and we do not know whether the deck has been reshuffled. As we will discuss, whether to assume that events are independent is crucial in assessing natural hazards. It makes a major difference whether or not we assume that a big earthquake, volcanic eruption, flood, or hurricane in an area makes the probability of another in the near future larger, smaller, or the same.

4.3   The Gaussian Distribution

The probabilistic approach views the actual outcome as a sample from a parent distribution, a population of all possible outcomes. The probability of each outcome is described by a *probability density function*, or *pdf*.

For example, we assumed that all the outcomes of rolling a die were equally likely, so each had a probability of 1/6. However, the only way to really know the probability density function is to actually do the experiment. After enough rolls, we have a good idea whether the die is fair. However, for natural hazards we do not know the probability density functions. This is like taking a new die from a package – we can assume that we know its probability density function, but will not know whether our assumption was right until we have rolled it many times. This problem occurs throughout science, but is especially a problem for natural hazards. In many other applications researchers can do experiments, but we have to wait for nature to provide the data, which can take thousands of years.

In science, the most commonly used probability distribution is the normal, or *Gaussian*, distribution, in which the probability of obtaining a certain value is greatest for the mean of the distribution and decreases for values on either side of it. This distribution (or probability density function) is visually repre-

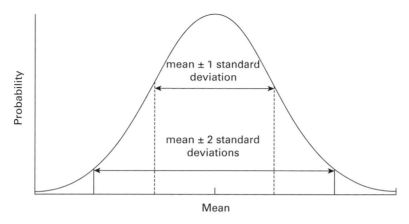

Figure 4.1 Probability of observing specific values from a Gaussian parent distribution.

sented by the famous bell-shaped curve (Figure 4.1), and is mathematically described by two parameters, its *mean* μ and *standard deviation* σ:

$$p(x) = \frac{1}{\sigma(2\pi)^{1/2}} \exp\left[\frac{-1}{2}\left(\frac{x-\mu}{\sigma}\right)^2\right]. \tag{4.26}$$

This expression gives the probability that a sample will come out in the range $x \pm dx$. The most probable value is the mean, and values on either side are less likely the further from the mean they are. The standard distribution or its square σ^2, known as the *variance*, measure the spread about the mean.

This distribution is often described using the normalized variable $z = (x - \mu)/\sigma$, which describes how far x is from its mean in terms of the standard deviation:

$$p(z) = \frac{1}{(2\pi)^{1/2}} \exp(-z^2/2). \tag{4.27}$$

To compute the probability $A(z)$ that a value is within a range z of the mean, we integrate the probability density function from $-z$ to z to find

$$A(z) = \int_{-z}^{z} p(y)dy = \frac{1}{(2\pi)^{1/2}} \int_{-z}^{z} \exp[-y^2/2]dy. \tag{4.28}$$

For $z = 1$, $A(1) = 0.68$, so there is a 68% percent chance that any measurement will be within a range of one standard deviation – above or below – the mean. Similarly, $A(2) = 0.95$ and $A(3) = 0.997$, so there are 95% and 99.7% chances that any measurement will be within two and three standard deviations of the mean.

This discussion leads to points that are important for studying natural hazards. First, the quantities that interest us, such as how often a flood of a certain size occurs, are estimated from a limited set of measurements. We can think of each measurement x_i as a sample from a parent distribution whose mean is the value we want to measure, and whose standard deviation reflects the uncertainty in each measurement. We expect that as the number of measurements gets large, their mean

$$\mu' = \frac{1}{N}\sum_{i=1}^{N} x_i = \frac{1}{N}(x_1 + x_2 + x_3 + \ldots + x_N) \qquad (4.29)$$

will approach the mean of the parent distribution and the spread of the measurements about the sample mean

$$\frac{1}{N}\sum_{i=1}^{N}(x_i - \mu')^2. \qquad (4.30)$$

will approach the parent distribution's variance σ^2. In reality N is often not that large, so the mean of the measurements, μ', known as the *sample mean*, can be different from μ, the parent distribution mean. An important result, which we do not prove here, is that if all the measurements have the same standard deviation σ, then the standard deviation of the sample mean is

$$\sigma_{\mu'} = \sigma/\sqrt{N}. \qquad (4.31)$$

This *standard deviation of the mean* is a measure of the uncertainty of the sample mean, which we use as our estimate of the quantity of interest. Thus our estimate of the sample mean is better – has smaller uncertainty – than the individual measurements. When we have only a small number of measurements, the sample mean has a large uncertainty. As we get more measurements, the uncertainty decreases. Making nine measurements of a quantity and averaging them gives a value three times more precise than each individual measurement.

The idea that taking more samples gives a better view of the parent distribution makes sense, because if we sampled the whole population, the sample mean and parent mean would be the same. However, when the number of samples is small, their mean can be very different from the population mean.

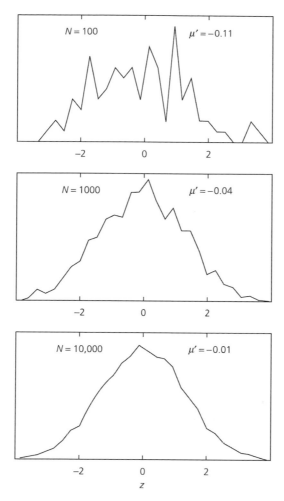

Figure 4.2   Histogram of the results of drawing N samples from a Gaussian parent distribution with mean zero and a unit standard deviation. (Stein and Wysession, 2003. Reproduced with permission of John Wiley & Sons.)

Figure 4.2 compares the results of taking different numbers of samples from a Gaussian parent distribution with a mean of zero and standard deviation of one. For small numbers of samples the observed distribution looks quite different from the parent distribution and the sample mean μ' differs from that of the parent distribution. As the number of samples increases, the observed distribution looks increasingly like the parent distribution. This effect is a problem in studying the recurrence of earthquakes or other natural hazard events, where the few samples available make it hard to assess what parent distributions and parameters should be used to estimate probabilities. In particular, the longer a time history available, the better we can estimate the mean recurrence time of an event.

As we discussed in section 3.3, people often claim to have observed unusual behavior in the earth and interpreted it as evidence for an upcoming event. For example, ground deformation, changes in seismic velocity, gas emissions, and other effects have been interpreted as precursors to earthquakes. An important question is when to regard such observations as meaningful.

Typically, we assume that observations are samples from a Gaussian parent distribution. If we observe some quantity and it differs by two standard deviations from the expected value, we could say that there was only a 5% risk that this difference resulted purely from chance. Thus the observation is said to be statistically significant at the 95% level. There are many pitfalls involved, including the fact that this interpretation depends on the assumed standard deviation. If the measurements are less accurate than we think, the standard deviation is greater and the observations are less significant.

Moreover, some observations will appear significant purely by chance. If the standard deviation is estimated correctly, 5% of the observations will appear significant at the 95% level, and 3% will appear significant at the 97% level. Given enough observations, some will seem significant.

For example, imagine that we are trying to decide if the ground in an area is deforming due to strain building up for a future earthquake. If we measure the distances between 20 GPS receivers, there are $20 \times 19 = 380$ baselines between different pairs of receivers. Thus even if the ground is not really deforming, 19 baselines (5%) will appear to show motion different from zero at the 95% confidence level, and 11 (3%) will look significant at 97%. If these apparent motions are simply due to chance, they will get smaller as later and more precise measurements – with smaller standard deviations – are made. Figure 4.3 illustrates this for GPS measurements in and across the New Madrid seismic zone in the Central US. As the precision of the measurements improves due to longer intervals of measurements, the maximum possible motion has shrunk rapidly from the large values originally reported to very small numbers.

4.4   Probability vs Statistics

Assessments of natural hazards depend significantly on the probability density functions that are assumed to describe the occurrence of events. Various distributions, including Gaussian ones, are used. Choosing which to use brings

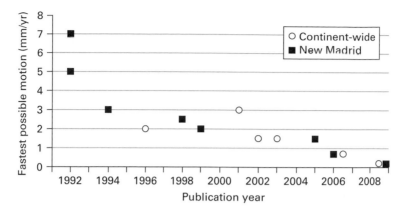

Figure 4.3 Maximum rate of motion in and across the New Madrid seismic zone shown by successively more precise GPS measurements. (Calais and Stein, 2009. Reproduced with permission of The American Association for the Advancement of Science.)

out the distinction between the branches of mathematics that deal with random variables. *Probability* deals with the outcomes of random processes whose probability density functions (pdfs) are known. For example, if we assume a coin is "fair" – equally likely to give a head or tail when flipped – we can calculate the probability of getting eight heads in ten flips. *Statistics* goes the other way, starting with a set of observations, characterizing them, and trying to figure out something about the process that gave rise to them. An example would be taking the results of a set of coin flips and trying to decide if the coin is fair, such that heads and tails are equally likely. If it is not fair, the question is how it is biased, and thus what pdf it gives rise to.

To look into this question, we use two key results from probability theory, the *Law of Large Numbers* and the *Central Limit Theorem*.

The *Law of Large Numbers* says that as the number of samples grows, the mean (average) of the samples converges to the mean of the parent population being sampled. Because we do not know the mean of the parent population, our best estimate of it is the mean of our samples, and this estimate will get better as the number of samples grows, as we have already discussed.

The *Central Limit Theorem* says that the mean of a sum of many independent random variables will be described by a normal distribution, even if the individual random variables are not described by normal distributions. This theorem explains why we use normal distributions to describe many physical processes. We think of the process as the net effect of many random elements, so the outcome can be viewed as their sum.

In this case, we define the outcome of the coin flip by a random variable x, with $x = 1$ when the coin comes up heads and $x = 0$ when it comes up tails. Since we do not know if the coin is fair, we call the probability of a head p, where $1 > p > 0$, and the probability of a tail q, where $q = 1 - p$. For any given flip, the expected value of x is the sum of the two outcomes weighted by their probabilities:

$$E(x) = 1 \times p + 0 \times q = p. \tag{4.32}$$

The variance of x is the sum of the squared difference between each outcome and the expected value, weighted by their probabilities

$$\sigma^2 = (1 - p)^2 p + (0 - p)^2 (1 - p) = pq. \tag{4.33}$$

If we draw n independent samples, their expected sum is $nE(x) = np$ and their variance is $n\sigma^2 = npq$. Therefore if the coin is fair, with $p = \frac{1}{2}$, the expected number of heads in a sample of $n = 100$ is $np = 100(1/2) = 50$. The variance should be $n\sigma^2 = (100)(1/4) = 25$, so the standard deviation should be $\sigma = 5$.

Imagine a case where we get 35 heads in 100 flips. This seems very low, so we suspect bias. The deviation between the actual number of heads and that expected from a population where $p = \frac{1}{2}$ is $|35 - 50| = 15$, which is $15/5 = 3$ standard deviations. By the Central Limit Theorem, we can consider the sum of the 100 coin flips to be a normally distributed variable, so the probability of being three standard deviations from the mean is only 0.003. Because such an outcome is quite unlikely, the coin is probably biased.

4.5   Shallow and Deep Uncertainties

For natural hazards, we take an analogous approach to that described in the previous section. Because we do not know the probability density functions or their parameters, we start with the known history of events, add our ideas about how the world works, and try to infer a probability distribution that describes that history. Hopefully, the pdf we infer will do a reasonable job of predicting the probability of future events. Sometimes this works well, and often it doesn't.

In doing this, we are trying to characterize unknown future events, with limited knowledge. In a seminal paper, economist Frank Knight (1921) proposed that to distinguish between "the measurable uncertainty and an unmeasurable one, we may use the term 'risk' to designate the former and the term 'uncertainty' for the latter." (This terminology is different from that in the natural hazards literature, where "hazard" denotes the natural occurrence of phenomena, and "risk" denotes the dangers that hazards pose to lives and property.)

Since Knight's paper, various terms and ways of classifying uncertainty have been used in different fields. Seismic hazard analysis, which was developed by engineers, follows the engineering literature in distinguishing uncertainties by their sources. Those due to irreducible physical variability of a system are called *aleatory*. The term comes from the Latin word for dice, "aleae". (In 49 BC Julius Caesar famously said "alea iacta est" – *"the die has been cast"* – when he led his army across the Rubicon River into Italy, putting himself in rebellion against the Roman government.) In contrast, uncertainties are called *epistemic* uncertainties when they are due to lack of knowledge of the physical system.

Another way to classify uncertainty is by whether we have a good way to describe it. *Shallow uncertainties* arise when our estimate of the probability density function is reasonably good, so the probabilities of outcomes are reasonably well known. In such cases, past events are good predictors of future ones. For example, flipping a fair coin should yield either a head or tail, each with a probability of ½. If this model and parameter describe the results of past flips well, we have high confidence in their ability to predict future flips.

In contrast, *deep uncertainties* arise when the probabilities of outcomes are poorly known, unknown, or unknowable. This occurs when we have multiple possible models with poorly known parameters, either because we do not adequately understand the system or it has inherently unpredictable elements. In such situations, past events may give little insight into future ones.

For example, whether a baseball player coming to bat will get a hit is described by shallow uncertainty, because his batting average is a good predictor of this probability. In contrast, deep uncertainty arises in trying to predict the winner of the World Series in the next baseball season. The teams' past performances provide only limited insight into the future of a complicated process. We could develop various models based on the past performance, but would have little confidence in them.

As we will see, both types of uncertainties arise for natural hazards. We treat the occurrence of floods or hurricanes as involving shallow uncertainties, so a standard probability density function is used, with parameters estimated reasonably well from the data. These are modeled as time-independent events, assuming that the history of events gives a reasonable estimate of their future probability, which improves as the history available gets longer. However, forecasts with a long timescale face deep uncertainties associated with possible effects of climate change, because rainfall patterns and storm frequencies or intensities may change in ways that are hard to predict from climate models.

Such changes seem likely to be occurring already. Following the effects of Hurricane Sandy in October 2012, which caused damage estimated at more than $60 billion to the northeastern US, New York governor Andrew Cuomo said "I told President Obama, we have a 100-year flood every two years . . . the frequency of extreme weather is going way up." Less than a year later, major floods in Central Europe did enormous damage. A cafe owner in Pirna, Germany, who was trying to keep the highest flood water in five centuries out of his cafe, complained, "The flood of a century is supposed to happen once in a lifetime, not once every 10 years." Insurance company climatologists explained that the flood resulted from a "weather pattern that has become increasingly common in recent decades." Winter storm tracks from the Atlantic are shifting northward, causing increased rain and flooding in northern Europe and increasing drought in southern Europe.

Deep uncertainties also arise in estimating the occurrence of earthquakes, because appropriate models for where and how often they recur or how big they will be are not well known, despite decades of study.

As we have seen already and will discuss further, natural hazard assessments and thus mitigation policies often fail because the hazard assessment underestimated the deep uncertainties. Recognizing these uncertainties is crucial for doing better.

Questions

4.1.   How would the expected annual earthquake loss in equation 4.7 change if magnitude 5 earthquakes occurred once every 20 years? How would it change if magnitude 7 earthquakes occurred every 500 years?

4.2.   Calculate the expected value of betting on a single number in roulette, given that if the number comes up the player wins 35 times the sum bet. How would this change for a European roulette wheel that has 0 but not 00?

4.3.   Earthquake hazards are often described by the maximum shaking expected on average once in 500 or 2500 years. Find the probability that during a fifty-year period, the life of a typical building, the building will experience these levels of shaking.

4.4.   In 2003, a magnitude 6.6 earthquake destroyed much of the city of Bam, Iran. Much of the city, including its famous 2000-year-old citadel, was made of mud brick that is very vulnerable to earthquake shaking and thus was seriously damaged. What insights does the city's history give for how often on average such earthquakes happen? What can you say and not say?

4.5.   Find the conditional probability of drawing a heart from a deck of cards after one card, which was not a heart, was drawn and not replaced. Also find the joint probability of these two events.

4.6.   Card-counting blackjack players can be defeated if the casino dealer shuffles the cards after every round. Should casinos do this all the time, rather than only if they suspect card-counting? Why or why not?

4.7.   Problem 1.6 explains how to calculate the odds of winning a state lottery. Explain this calculation using the terminology in this chapter. What is the expected value of buying a ticket? When it is rational to play?

4.8.   Use a computer program or an Excel spreadsheet to compute the Gaussian probability density function (equation 4.26). There is more than one way to do this, including by calculating the exponential or by using Excel's NORMDIST function, where $p(z) =$ NORMDIST $(z,0,1,\text{FALSE})$. Use the result to plot a curve like Figure 4.1.

4.9.   NORMDIST or an equivalent program can also be used to find the probability that a value is within a range z of the mean (equation 4.28). In Excel, this is done by using NORMDIST$(z,0,1,\text{TRUE})$ – NORMDIST$(-z,0,1,\text{TRUE})$. Test that you get the correct values for $z = 1$, 2, and 3.

4.10.   How would you try to decide if the apparent ground motion in a GPS survey like that described in section 4.3 is real or appears purely by chance?

4.11.   A challenge for medical researchers involves "cancer clusters," communities that have higher-than-expected cancer rates. Some clusters would be expected to occur purely from chance, because of 100 communities, 5% should have rates that are significantly higher at 95% confidence. How could one try to identify which clusters result from factors other than chance?

4.12.   Verify the result in equation 4.33.

Further Reading and Sources

The epigraph is from Feynman (2000).

Many popular treatments of probability including Taleb (2004) and Mlodinow (2008) discuss issues including whether an apparent pattern is real or random. Silver's (2012) discussions include some geophysical cases. Popular treatments of statistics include Huff (1993) and Best (2001, 2004), who considers their role in forming policy, often incorrectly. Bevington and Robinson (1992), Young (1996), and Taylor (1997) are good introductions to probability and statistics for scientists. The concept of deep uncertainty is discussed by Morgan et al. (2009), Cox (2012), Hallegatte et al. (2012), and Stein and Stein (2013).

Cuomo's comment is reported by Dwyer (2012). Eddy (2013a,b) describes the Central European floods.

References

Best, J., *Damned Lies and Statistics: Untangling Numbers from the Media, Politicians, and Activists*, University of California, Berkeley, CA, 2001.

Best, J., *More Damned Lies and Statistics: How Numbers Confuse Public Issues*, University of California, Berkeley, CA, 2004.

Bevington, P. R., and D. K. Robinson, *Data Reduction and Error Analysis for the Physical Sciences*, Second Edition, McGraw-Hill, New York, 1992.

Calais, E., and S. Stein, Time-variable deformation in the New Madrid seismic zone, *Science*, *5920*, 1442, 2009.

Cox, L. A., Jr., Confronting deep uncertainties in risk analysis, *Risk Anal.*, *32*, 1607–1629, 2012.

Dwyer, J., Reckoning with realities never envisioned by city's founders, *New York Times*, October 30, 2012.

Eddy, M., Merkel visits flood-stricken regions of Germany, *New York Times*, June 4, 2013a.

Eddy, M., In flooded Europe, familiar feelings and new questions, *New York Times*, June 6, 2013b.

Feynman, R., *The Pleasure of Finding Things Out*, Basic Books, New York, 2000.

Hallegatte, S., Shah, A., Lempert, R., Brown, C., Gill, S., Investment Decision Making Under Deep Uncertainty – Application to Climate Change. Policy Research Working Paper 6193, The World Bank, Sustainable Development Network Office of the Chief Economist, September 2012.

Huff, D., *How to Lie with Statistics*, W. W. Norton, New York, 1993.

Knight, F., *Risk, Uncertainty, and Profit*, Houghton Mifflin, Boston, MA, 1921.

Mlodinow, L., *The Drunkard's Walk: How Randomness Rules Our Lives*, Pantheon, New York, 2008.

Morgan, G. M., Dowlatabadi, H., H. Henrion, H. et al. *Synthesis and Assessment Product 5.2 Report by the U.S. Climate Change Science Program and the Subcommittee on Global Change Research*, The National Academies Press, Washington, DC, 2009.

Silver, N., *The Signal and the Noise*, Penguin, New York, 2012.

Stein, S., and J. L. Stein, Shallow versus deep uncertainties in natural hazard assessments, *Eos Trans. AGU*, *94*, 133–134, 2013.

Stein, S., and M. Wysession, *Introduction to Seismology, Earthquakes, and Earth Structure*, Blackwell, Oxford, 2003.

Taleb, N. N., *Fooled By Randomness*, Random House, New York, 2004.

Taylor, J. R., *An Introduction to Error Analysis: The Study of Uncertainties in Physical Measurements*, University Science Books, Sausalito, CA, 1997.

Young, H., *Statistical Treatment of Experimental Data*, Waveland, Long Grove, IL, 1996.

5

Communicating What We Know and What We Don't

When in doubt tell the truth. It will confound your enemies and astound your friends.

Mark Twain

5.1 Recognizing and Admitting Uncertainties

For natural hazard forecasts to be fully useful, people need to know how much confidence to put in them. However, as we have seen, although scientists know a lot about hazards, there is also a lot that they do not know. As seismologist Hiroo Kanamori (2011) noted:

> the 2011 Tohoku earthquake caught most seismologists by surprise . . . even if we understand how such a big earthquake can happen, because of the nature of the process involved we cannot make definitive statements about when it will happen, or how large it could be.

Thus while scientists try to learn more about the earth and do better at forecasting hazards, it is important to maintain humility in the face of the complexities of nature. We need to be honest with ourselves and the public about the present limits of knowledge. It is important to explain these limits when planning for possible future disasters, where the consequences of overestimating or underestimating the hazard can be high.

Playing against Nature: Integrating Science and Economics to Mitigate Natural Hazards in an Uncertain World, First Edition. Seth Stein and Jerome Stein.
© 2014 John Wiley & Sons, Ltd. Published 2014 by John Wiley & Sons, Ltd.
Companion Website: www.wiley.com/go/stein/nature

A good approach is General Colin Powell's (2012) charge to intelligence officers: "Tell me what you know. Tell me what you don't know. Then tell me what you think. Always distinguish which is which." Had Powell followed this approach as US Secretary of State, he would not have championed the 2003 invasion of Iraq, most famously in an inaccurate speech to the United Nations. As he admitted afterwards, "a blot, a failure will always be attached to me and my UN presentation."

In dealing with natural hazards, Powell's last point can be modified to: "Always distinguish which is which *and why*," so that people understand where the uncertainty comes from.

Among natural hazard forecasters, meteorologists lead in explaining the uncertainties in their forecasts to the public. One approach is to point out that different predictions can be made by different groups using different assumptions. For example, on February 2, 2000 the *Chicago Tribune's* weather page stated:

> Weather offices from downstate Illinois to Ohio advised residents of the potential for accumulating snow beginning next Friday. But forecasters were careful to communicate a degree of uncertainty on the storm's precise track, which is crucial in determining how much and where the heaviest snow will fall. Variations in predicted storm tracks occur in part because different computer models can infer upper winds and temperatures over the relatively data-sparse open Pacific differently. Studies suggest that examining a group of projected paths and storm intensities – rather than just one – helps reduce forecast errors.

Graphics in the paper compared four predicted storm tracks and seven precipitation estimates for Chicago.

In the Powell formulation, "you know" that a storm is coming, "you don't know" its exact track and thus how much snow will fall where – illustrated by the comparison of the varying model predictions, "you think" that snow accumulation is likely, and "which is which and why" are the models' uncertainties and their limitations, in part due to sparse data.

Hurricane forecasts also give uncertainties. For example, in December 2008, once Hurricane Ike had formed, the US National Weather Service predicted that it would continue westward, and then turn north along the Florida coast. This prediction is shown by the shaded area on the forecast map, which was assumed to have a 95% probability of containing the storm's track (Figure 5.1). This area widens along the predicted track because the uncertainty increases for longer times.

The area at risk prepared for the oncoming storm; this included canceling an academic program review meeting at the University of Miami school of Marine and Atmospheric Sciences that the first author was to attend. However, the storm ignored the forecast and headed into the Gulf of Mexico, striking the Texas coast.

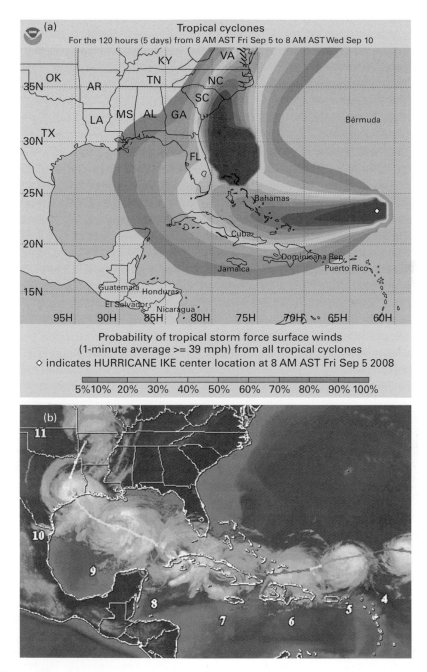

Figure 5.1   Comparison of the predicted (a) and actual (b) tracks of Hurricane Ike. ((a) NOAA, 2008. (b) NOAA, 2009.)

Thus even forecasts with stated uncertainties can have problems. As for the Tohoku earthquake, the question is whether there was something inherently wrong with the forecast model or whether it was just bad luck that a low probability event happened. After all, 5% of the time the hurricane's actual track should be outside the forecast's 95% confidence level.

An example of good communication is the way, as Hurricane Irene threatened the US East Coast in August 2011, meteorologist Kerry Emanuel explained that:

> We do not know for sure whether Irene will make landfall in the Carolinas, on Long Island, or in New England, or stay far enough offshore to deliver little more than a windy, rainy day to East Coast residents. Nor do we have better than a passing ability to forecast how strong Irene will get. In spite of decades of research and greatly improved observations and computer models, our skill in forecasting hurricane strength is little better than it was decades ago.

The 2007 report by the Intergovernmental Panel on Climate Change nicely compares predictions of the effects of global warming due to carbon dioxide buildup in the atmosphere. Results of different models developed by different groups using different methods and assumptions are presented and discussed. The report compares the predictions of 18 models for the expected rise in global temperature. It further notes that the models "cannot sample the full range of possible warming, in particular because they do not include uncertainties in the carbon cycle. In addition to the range derived directly from the multi-model ensemble, Figure 5.2 depicts additional uncertainty estimates

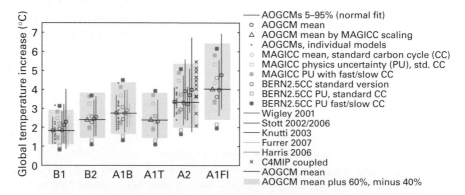

Figure 5.2   Comparison of the rise in global temperature by the year 2099 predicted by various climate models. For various scenarios of carbon emissions – B1, B2, etc. – the vertical band shows the different predicted warming. (IPCC, 2007. Reproduced from IPCC and Cambridge University Press.)

obtained from published probabilistic methods using different types of models and observational constraints." Thus although it would not be useful to say that the temperature will rise by 3.6 degrees, it is sensible to say that if these models describe the climate system adequately, the temperature is likely to rise by between one and seven degrees. This range is smaller for a specific scenario of carbon emissions. In the Powell formulation, the report says "we know the earth is warming, we do not know how fast because of the complicated climate system, we think it will be in this range."

These examples bring out how crucial it is to present uncertainties in forecasts. It is especially important to present what we think the uncertainties are, because scientists have learned the hard way that uncertainties are usually even larger.

A famous example is the number of chromosomes in the human body. For thirty years, researchers using microscope photos to count the chromosomes (DNA strands) in human cells were convinced there were $48:23$ pairs plus the X and Y chromosomes that determine sex. In 1956, better analysis methods showed that there were only 46. For years, researchers had followed the "consensus" that everyone "knew" was right, even though many probably first came up with another answer. Because people are involved, this happens much more often in science than ideally it should. Eventually, because nature is not affected by human consensus, the bandwagon topples.

How this happens is shown by the history of measurements of the speed of light (Figure 5.3). Because this speed is crucial for the theory of special relativity, it is one of the most frequently and carefully measured quantities in science. From 1875 to 1900, all the experiments found speeds that were too high. Then, from 1900 to 1950, the opposite happened – almost all the experiments found speeds that were too low. This kind of error, where results are generally on one side of the real value, is called a *bias*. It probably happened because over time, experimenters adjusted their results to match what they expected to find. If a result fit what they expected, they kept it. If a result did not fit, they threw it out. They were not being intentionally dishonest – just influenced by the conventional wisdom. The pattern, called a *bandwagon effect*, only changed when someone had the courage to report what was actually measured, not what was expected. Experience in many branches of science shows that bandwagon effects often happen.

This human tendency to see what we expect to see, or accept data that fit our view and ignore data that do not, is called *confirmation bias*. It has been recognized since AD 1620, when Francis Bacon pointed out that

the human understanding, once it has adopted an opinion, collects any instances that confirm it, and though the contrary instances may be more numerous and

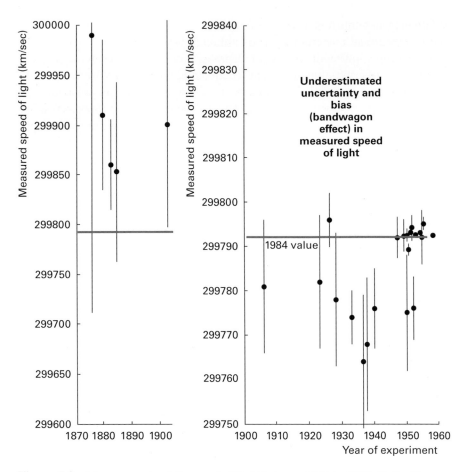

Figure 5.3   Measurements of the speed of light between 1875 to 1960. Vertical bars show the experimenters' assessments of the uncertainty in their measurements. (Henrion and Fischhoff, 1986. Reproduced with permission of the American Association of Physics Teachers.)

more weighty, it either does not notice them or else rejects them, in order that this opinion will remain unshaken.

It amounts to fooling ourselves, but is hard to avoid. As a saying goes, "It's not what you don't know that hurts you – it's what you know that isn't so."

Confirmation bias is a major problem in science. As Richard Feynman warned, "The first principle is that you must not fool yourself – and you are the easiest person to fool." To make sense of the complicated world, scientists rely on their instincts about how it works, and use them to filter out "bad" data. When this works, progress results. However, it makes it natural to plan research to confirm what is expected rather than look for what is not, and interpret the results accordingly. Similarly, key data are often missed or ignored because scientists tend to discount or not see what they do not expect. A famous example is an experiment where people are asked to watch a video of a basketball game and count how many times one team gets the ball. Concentrating on the count, viewers often miss someone in a gorilla suit walking across the court. In hindsight, it often is hard to believe that so many researchers missed something important.

These effects are common in natural hazard assessment and mitigation situations. Given the major economic issues at stake, those involved with hazard assessments are reluctant to accept evidence showing flaws in the assessment. A prime example, discussed in Chapter 2, is the unwillingness of the Japanese authorities to rethink their assessment of the Tohoku tsunami hazard and the vulnerability of the Fukushima nuclear plant, even as new data accumulated.

5.2 Precision and Accuracy

The speed of light measurements also illustrate the common problem that uncertainties are generally underestimated. The experimenters thought their measurements were more precise than they really were. This happens in many applications, because estimating uncertainties is difficult.

We think about uncertainty using two concepts, *precision* and *accuracy*. To see their relation, imagine shooting at a target. Ideally, the shots will group closely around the bull's-eye (Figure 5.4a). These shots are very precise, in that they are close to each other. They are also very accurate, in that they are near the bull's-eye. However, three other outcomes can happen. The shots could be precise – closely grouped – but inaccurate – away from the bull's-eye (Figure 5.4b). They could be imprecise – scattered widely – but accurate in that their average is close to the bull's-eye (Figure 5.4c). Worst of all, they could be both imprecise and inaccurate (Figure 5.4d).

These ideas are important in thinking about measurements of a physical quantity. Precision describes how repeatable the measurements are. This is also called the *random error* in the measurements – how close we find the values to be when we measure the same thing many times. Accuracy, how close the measurements are to the "true" value of the quantity, is described by the *systematic error*. In the target analogy, a skilled shooter's shots will group tightly – precisely – but would miss the bull's-eye if the gunsight were inaccurate. A less skilled shooter with an accurate gunsight would produce shots that were more scattered – imprecise – but with no bias in any particular direction relative to the bull's-eye.

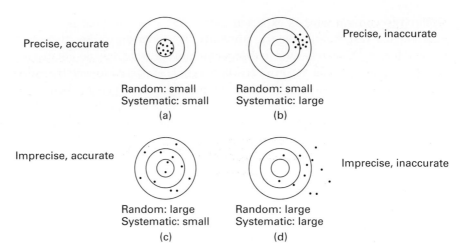

Figure 5.4   Illustration of the concepts of precision and accuracy. (Taylor, 1997. © University Science Books, Mill Valley, CA. Used with permission. All rights reserved.)

In the real world, precision is much easier to assess than accuracy. The problem is that we do not know the true value of what we are trying to measure. If we knew it, we would not be measuring it. In the target analogy, we do not know where the bull's-eye is. We know the precision – how closely the shots group – but cannot say how close that group is to the bull's-eye.

We try to get around this problem by using different measurement systems, assuming that each system has a different systematic error. For example, measuring the length of a table many times with a meter stick gives an estimate of the precision – repeatability – of that measurement, and comparing the result to multiple measurements with a tape measure gives an idea of the accuracy.

These issues are usually addressed using statistical methods. The actual values measured are viewed as samples drawn from a parent distribution, a population of all possible measurements. Typically, the parent distribution is assumed to be a Gaussian or normal distribution, as described in section 4.3. For this distribution, 68% of the measurements should be within a range of one standard deviation – above or below – the mean. However, for the speed of light measurements, only about 50% of the error bars include the true value, rather than the 68% that should. This means that the researchers underestimated their uncertainties – the error bounds should be greater. This turns out to be a common, serious, and hard-to-avoid problem.

5.3 Testing Forecasts

As we have just seen, uncertainties are hard to estimate when measuring a physical quantity. They are even harder to estimate for forecasts of natural events, which are statements about the future of systems that in many cases we do not understand well.

To make things worse, we are especially interested in the probabilities of the most dangerous, and hence the rarest events. The rarer something is, the harder it is to estimate its probability. It is a lot easier to estimate the probability that a house will have a fire than the probability that it will be hit by a meteorite.

As noted in section 4.5, forecasters speak of *aleatory* (from the Latin word for dice) uncertainties as those due to random uncertainties in their model, and *epistemic* uncertainties due to systematic errors. Most models contain estimates of these quantities, but deciding how good these estimates are is tough. The only meaningful way to do it is to compare the model's predictions to what actually happened.

Nature often shows us that a model was not very good. An easy way to see this is that the speed-of-light example shows up in discussions of the failures of both of earthquake hazard maps and river flood predictions. The hazard models seemed sensible, but did not work very well. In hindsight, the uncertainties turned out to be much larger than had been thought. There is debate about whether this will turn out to be true for the global warming forecasts.

An often-quoted formulation of the idea that the only real test of models is how they describe data comes from Richard Feynman (1986). He explained that scientists have learned that

> whether they like a theory or they do not like a theory is not the essential question. Rather, it is whether or not the theory gives predictions that agree with the experiment. It is not a question of whether a theory is philosophically delightful, or easy to understand, or perfectly reasonable from the point of view of common sense.

In this spirit, it is important to test models to see whether their predictions are better than predictions based on null hypotheses, which are usually based on random chance. If not, then we should not use them, regardless of how sensible they seem, how much we like them, or how many people have been using them. Objective testing gets us past the confirmation bias, where people place more weight on data that fit their view about how the world works than on data that do not fit.

This kind of testing often is not done, because the results can be embarrassing. However, it is crucial in deciding how much faith to put in hazard models when making billion dollar decisions.

An analogy is the way that medical researchers are starting to use a new approach, called evidence-based medicine, which objectively evaluates widely used treatments. The results are often not what doctors and hospitals who have been using these treatments want to hear. Many widely prescribed and expensive medicines turn out to work no better than placebos – inert sugar pills. Even common surgery sometimes has this problem. Although more than 650,000 arthroscopic knee surgeries at a cost of roughly $5,000 each were being performed each year, a controlled experiment showed that "the outcomes were no better than a placebo procedure."

The best-tested natural hazard forecasts are weather forecasts, which are routinely evaluated to assess how well their predictions matched what actually occurred. A key part of this assessment is adopting agreed criteria for "good" and "bad" forecasts. Murphy (1993) notes that "it is difficult to establish well-defined goals for any project designed to enhance forecasting performance without an unambiguous definition of what constitutes a good forecast." Forecasts are also tested against various null hypotheses, including seeing if they do better than using the average of that date in previous years, or assuming that today's weather will be the same as yesterday's. Over the years, this process has produced measurable improvements in forecasting methods and results, and yielded much better assessment of uncertainties.

Similar approaches are starting to be used for other hazard forecasts. The fact that earthquake hazard maps often fail points out the need to test them and establish how well they work. A major question is whether these failures are just bad luck, which would be expected some of the time because the maps give probabilities, or whether they indicate problems with the maps. The latter seems more likely, because remaking maps after an earthquake implicitly indicates a problem with the map. If a probability model is correct, it does not need to be changed after a rare event. When someone wins the lottery, the commission does not change the odds.

Sorting this out requires objective criteria for testing maps by comparing their predictions to the shaking that actually occurred after they were published. Such testing would show how well the maps worked, give a much better assessment of their true uncertainties, and indicate whether or not changes in map-making methods over time give maps that work better.

We want maps that neither overpredict nor underpredict what happens. Underprediction leads to more damage than expected, and overprediction diverts resources that could have been used better for some other purpose and hurts economic growth. Thus we want to test maps with various metrics that reflect both overpredictions and underpredictions. We need algorithms that are not biased toward underprediction, because in any short time window, predicting no shaking is a good bet. Similarly, we do not want to favor over-prediction, because predicting that shaking will be less than a very high level is generally also a good bet.

A natural test is to compare the maximum acceleration (shaking) observed over the years in regions within the hazard map to that predicted by the map and by some null hypotheses. This could be done in various ways. One way is to apply the "skill score" method used to test weather forecasts. It would consider a region divided into N subregions. In each subregion i, over some time interval, we would compare the maximum observed shaking x_i to the map's predicted maximum shaking p_i. We then define and compute the Hazard Map Error (HME) as follows:

$$HME(p, x) = \Sigma_i (x_i - p_i)^2 / N. \tag{5.1}$$

We can then assess the map's skill by comparing it to the misfit of a reference map produced using a null hypothesis

$$HME(r, x) = \Sigma_i (x_i - r_i)^2 / N \tag{5.2}$$

using the skill score (SS)

$$SS(p, r, x) = 1 - HME(p, x) / HME(r, x). \tag{5.3}$$

The skill score would be positive if the map's predictions did better than those of the map made with the null hypothesis, and negative if they did worse. We could then assess how well maps have done after a certain time, and whether successive generations of maps do better.

One simple null hypothesis is to assume that the earthquake hazard is the same over an area. Figure 2.1 suggests that the Japanese hazard map is per-forming worse than this null hypothesis. Another null hypothesis is to assume that all oceanic trenches have magnitude 9 earthquakes about equally often (there is about one every 20 years on a trench somewhere in the world).

The idea that a map including the full detail of what is known about an area's geology and earthquake history may not perform as well as assuming the hazard is uniform at first seems unlikely. However, it is not inconceivable. An analogy could be describing a function of time composed of a linear term plus a random component. A detailed polynomial fit to the past data describes them better than a simple linear fit, but can be a worse predictor of the future than the linear trend (Figure 5.5). This effect is known as

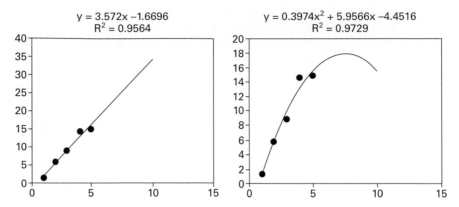

Figure 5.5　Illustration of overfitting by comparison of linear and quadratic fits to a set of data. The quadratic gives a better fit to the points but a poorer representation of the trend. (Stein et al., 2012. Reproduced with permission of Elsevier, B.V.)

overparameterization or *overfitting*. A test for this possibility would be to smooth hazard maps over progressively larger footprints. There may be an optimal level of smoothing that produces better performing maps.

A number of other issues arise in testing earthquake hazard maps or other natural hazard forecasts. This is harder than testing weather forecasts for several reasons. One is the longer timescales. Another is that because hazard assessments involve many subjective assumptions, we cannot develop a model using part of the data and test it with the rest. However, there are ways to test models with short datasets, including combining data from different areas. It is important to avoid the "Texas sharpshooting" bias (section 1.2) by using only data that were not used in making the hazard map.

5.4　Communicating Forecasts

Because the goal of hazard assessment is to help society develop sensible mitigation strategies, communicating both assessments and their uncertainties is crucial. Otherwise, most users will have no way to tell which forecasts are likely to be reasonably well constrained, and which are not. Having this information would help users make better decisions.

The need to communicate uncertainties is illustrated by events prior to the magnitude 6.3 earthquake that struck the area around the city of L'Aquila, Italy, in April 2009. The area is part of an active seismic zone along the Appenine mountain range that has had large earthquakes in the past and is recognized as hazardous (Figure 5.6). Thus when over a period of several

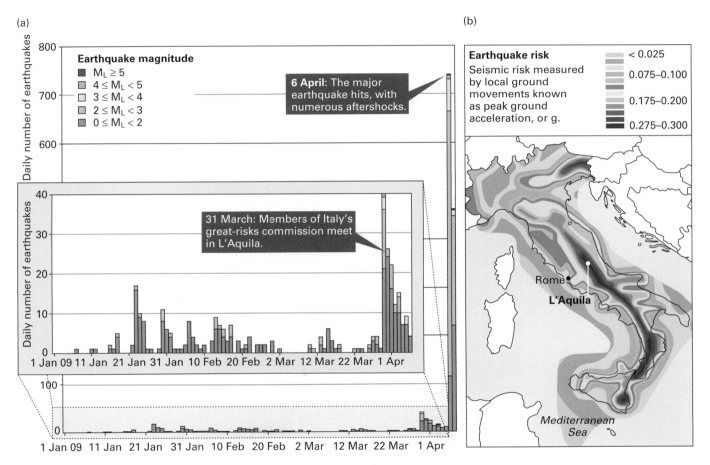

Figure 5.6 (a) Number of earthquakes and their magnitudes in the L'Aquila area, during the period leading up to the large April 6, 2009 earthquake. (b) Earthquake hazard map of Italy. (Hall, 2011. Reproduced with permission of Nature Publishing.)

months a swarm of hundreds of small earthquakes occurred, concern arose as to whether a larger earthquake was coming. The earthquake history showed that about 2% of such swarms were followed by larger earthquakes, whereas the others were not. Hence it would have been plausible to say that based on this record, the probability of a large earthquake seemed larger than before the swarm, but was still quite small. Although seismologists on a commission advising the authorities recognized this, the civil protection official on the commission told the public that the situation posed "no danger." When prompted by a journalist, "So we should have a nice glass of wine," the official replied "Absolutely." However, the earthquake that struck a week later caused over 300 deaths and triggered legal proceedings against the civil protection official and seismologists who advised him.

This tragic episode raised the question of what could have usefully been done with the information available. Traditionally, some residents would sleep outside their homes, which they knew were vulnerable, during periods of seismic activity. In this case, as one said, "That night, all the old people in L'Aquila, after the first shock, went outside and stayed outside for the rest of the night. Those of us who are used to using the Internet, television, science – we stayed inside." However, continuing to sleep outside would have become impractical if the low-level seismicity had gone on for more than a few days.

The argument is sometimes made that the public cannot understand or use information including uncertainties. In fact, people and businesses like weather forecasts that include probabilities and find them useful. More generally, people make decisions in uncertain situations all the time. People bet on sporting events, buy homes and insurance, save for college and retirement, and make all kinds of other decisions that involve uncertainty and probability. These involve deciding how to use their limited resources, given uncertainty about the future. They also know that "experts" like stockbrokers, financial planners, and insurance agents sometimes are right, and sometimes are not.

That said, we would like certainty, although we know we cannot have it. As a result, we often accept predictions without asking about their uncertainties. Astrology is a multimillion-dollar industry, although there is no evidence that it works. This desire is explored by Dan Gardner's book *Future Babble: Why expert predictions fail and why we believe them anyway.* He tells the story of Nobel Prize winner Kenneth Arrow, describing his experience as a military weather forecaster in World War II:

> My colleagues had the responsibility of preparing long-range weather forecasts, i.e., for the following month. The statisticians among us subjected these forecasts to verification and found they differed in no way from chance. The forecasters themselves were convinced and requested that the forecasts be discontinued.

> The reply read approximately like this: "The commanding general is well aware that the forecasts are no good. However, he needs them for planning purposes."

Obviously, we want to do better before making major policy decisions, and because similar questions arise in many applications, there has been considerable discussion of how best to make them. The most important general conclusion is that these decisions cannot be made for a community by "experts," because there really are none. Researchers who have thought about issues can tell what they know, do not know, and think – but no one can say with confidence what will happen. There may be widely used models – like the Japanese hazard map – but the fact that they are widely used does not mean that nature has to follow them. The models can be harmful in some cases, as Pilkey and Pilkey-Jarvis explain in their book *Useless Arithmetic: Why Environmental Scientists Cannot Predict the Future*.

> The reliance on mathematical models has done tangible damage to our society in many ways. Bureaucrats who do not understand the limitations of modeled predictions often use them . . . Agencies that depend on project approvals for their very survival (such as the US Army Corps of Engineers) can and frequently do find ways to adjust the model to come up with correct answers that will ensure project funding. Most damaging of all is the unquestioning acceptance of the models by the public because they are assured that the modeled predictions are the way to go.

Scientists need to discuss the issues, present the state of knowledge, and help society explore policy options. It is crucial to explain the limitations of scientific knowledge and the relevant models. Sarewitz et al. (2000) argue that the best approach is open discussion:

> Above all, users of predictions, along with other stakeholders in the prediction process, must question predictions. For this questioning to be effective, predictions must be as transparent as possible to the user. In particular, assumptions, model limitations, and weaknesses in input data should be forthrightly discussed. Institutional motives must be questioned and revealed . . . The prediction process must be open to external scrutiny.
>
> Openness is important for many reasons but perhaps the most interesting and least obvious is that the technical products of predictions are likely to be "better" – both more robust scientifically and more effectively integrated into the democratic process – when predictive research is subjected to the tough love of democratic discourse . . .
>
> Uncertainties must be clearly understood and articulated by scientists, so users understand their implications. If scientists do not understand the uncertainties – which is often the case – they must say so. Failure to understand and

articulate uncertainties contributes to poor decisions that undermine relations among scientists and policy makers.

A prime example of this problem occurred in 2008, as Hurricane Ike approached the Texas coast (Figure 5.1). In 1900, this area was the site of the deadliest hurricane in US history, which killed at least 6,000 people. As Ike approached, the National Weather Service warned that people who did not evacuate coastal communities faced "certain death" due to winds and especially the storm surge, the high water produced by storm winds. In fact, fewer than 50 of the 40,000 who stayed on Galveston Island were killed. The predicted 100% probability of death – with no uncertainty – proved at least 800 times too high, giving a new meaning to the old line about "close enough for government work." Of the residents who heard the "certain death" warning, about equal numbers had positive and negative responses. Positive responses included "blunt . . . effective," "correct," "to the point," "scared you to death," and "people who didn't heed were foolish." Negative responses included "harsh and overreactive," "overblown," "ridiculous," "humorous," "stupid," "rude," and "not appropriate." It seems that the "certain death" warning helped to convince some to evacuate, while making others who considered it "overly dramatic or not credible" less likely to respond to future warnings.

Thus although such exaggerated worst-case warnings may save lives, repeated overpredictions without acknowledging uncertainty can cause the public to ignore warnings. The same issue arises for tsunami warnings, as discussed in section 3.2. Hence instead of overhyping the hurricane threat, it would have been better to follow the Powell approach (section 5.1): "we know a huge storm is coming; we do not know how large the storm surge will be because that's hard to predict; we think it's very dangerous and people should evacuate."

A similar approach would be useful for earthquake hazards, which are traditionally presented by a map showing "the" hazard – the maximum shaking expected with some probability in some time interval. As we have seen, large destructive earthquakes often occur in areas predicted by the maps to be relatively safe. The problem is that these maps often have large uncertainties, because they depend on many poorly constrained parameters. Because these were not disclosed, the higher-than-predicted shaking came as a rude surprise.

It thus would be more useful to show a range of predictions. Figure 5.7 compares hazard predictions for two cities in the Central US's New Madrid seismic zone. For St. Louis, varying the model parameters produces a range of predictions from 20% of the acceleration of gravity to 50%, and for Memphis the range is even greater, from about 30% to over 90%.The predictions depend on the assumed magnitude ("M7" or "M8") of the largest earth-

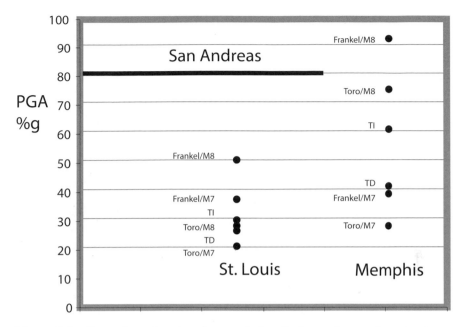

Figure 5.7   Comparison of earthquake hazard, described as peak ground acceleration (PGA) as a percentage of the acceleration of gravity expected with 2% risk in 50 years, predicted by various assumptions. (Stein et al., 2012. Reproduced with permission of Elsevier, B.V.)

quakes and the model chosen to predict ground shaking ("Frankel" or "Toro"). For example, the Frankel model predicts the highest hazard. The predicted hazard also depends on whether the probability of the largest earthquakes is assumed to be time-independent ("TI") or time-dependent ("TD") – small shortly after the past one and then increasing with time. Seeing this range of hazard estimates is helpful in considering whether it is worth spending billions of dollars to build buildings in the area to the high standards of earthquake resistance used in California.

Explaining the uncertainties involved with natural hazard assessments would help communities use these forecasts to make better policy. It is important to develop ways of characterizing these uncertainties in a rigorous sense. However, even if we cannot do that, there are ways to give a sense of the uncertainties. One is an "Italian flag" graphic, which shows a subjective assessment of the evidence for and against a proposition. Evidence for is shown by the green bar, evidence against by the red bar, and the uncertainty is shown as white. For example, in assessing the possibility that a magnitude 9 earthquake and megatsunami will occur along the Nankai Trough (Figure

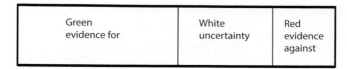

Green evidence for	White uncertainty	Red evidence against

Figure 5.8   Italian flag graphic showing an assessment of both the evidence for and against a proposition and the uncertainty involved, using the 50%, 30% and 20% percentages quoted in the text.

1.3), a seismologist might give 50% positive weight to the fact that such earthquakes have occurred on other subduction zones, 20% negative weight to the fact that none are known there from seismological or geological records, and 30% weight to the uncertainty resulting from factors including our limited knowledge of the subduction process and the limited paleotsunami record (Figure 5.8). Although this assessment is subjective, the flag format at least illustrates the estimated uncertainty.

Explaining and working with uncertainty in such situations remains a major research area both because of the scientific challenges involved and because natural hazard forecasts differ in some ways from those in daily life. Oreskes (2000) suggests recognizing that

> forecasts of events in the far future, or of rare events in the near future, are of scant value in generating scientific knowledge or testing existing scientific belief. They tell us very little about the legitimacy of the knowledge that generated them. Although scientists may be enthusiastic about generating such predictions, this in no way demonstrates their intellectual worth. There can be substantial social rewards for producing temporal predictions. This does not make such predictions bad, but it does make them a different sort of thing. If the value of predictions is primarily political or social rather than epistemic, then we may need to be excruciatingly explicit about the uncertainties in the theory or model that produced them, and acutely alert to the ways in which political pressures may influence us to falsely or selectively portray those uncertainties.
>
> As individuals, most of us intuitively understand uncertainty in minor matters. We do not expect weather forecasts to be perfect, and we know that friends are often late. But, ironically, we may fail to extend our intuitive skepticism to truly important matters. As a society, we seem to have an increasing expectation of accurate predictions about major social and environmental issues, like global warming or the time and place of the next major hurricane. But the bigger the prediction, the more ambitious it is in time, space, or the complexity of the system involved, the more opportunities there are for it to be wrong. If there is a general claim to be made here, it may be this: the more important the prediction, the more likely it is to be wrong.

Questions

5.1. How would you assess whether the fact that Hurricane Ike moved outside its forecast track (Figure 5.1) represented a problem with the forecast or was just chance?

5.2. Write a short public statement – less than 200 words – that you would have given to the public if you had been working for the National Weather Service as Hurricane Ike approached Galveston Island. Your goal is to realistically describe the situation and make sensible recommendations.

5.3. Write a short public statement – less than 200 words – that you would have given to the public if you had been working for the Italian civil protection authorities during the L'Aquila earthquake swarm. Given the public concern that a large earthquake may occur soon, your goal is to realistically describe the situation. The paper by Hall (2011) may be helpful.

5.4. Imagine that you are in charge of building an information display for a mountain resort town. One of the exhibit developers wants to include a multimedia display explaining how the beautiful scenery results from a volcanic eruption 6,000 years ago and how scientists monitor the area's continuing volcanic activity. Another argues that this may concern tourists and discourage people and businesses from moving to town. What would you do? How would you explain your decision to the town council?

5.5. The figure below shows schematically the results of fitting two models to a set of data. Which model describes the past data better? Which seems like a better prediction for the future? Which do you prefer and why?

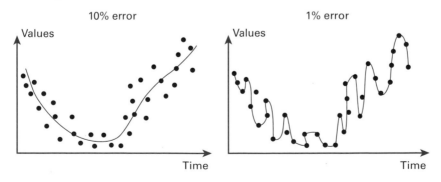

5.6. Drug company biologists who examined the results of important scientific studies about cancer have found that less than a quarter of the results can be replicated, as reviewed in "Trouble at the lab" (The Economist, October 19, 2013). What might be the causes of these problems?

Further Reading and Sources

Stirling (2010) discusses the need to present the true uncertainties in forecasts in policy-making applications. Morss et al. (2008) and Hirschberg et al. (2011) discuss communicating weather forecast uncertainties to the public.

Figure 5.1 is from *www.nhc.noaa.gov/archive/2008/graphics/al09/loop _PROB34.shtml* and *www.srh.noaa.gov/lix/?n=ike2008*.

Emanuel's article "Why are hurricane forecasts still so rough?", is from CNN, 25 Aug. 2011, available at *http://www.cnn.com/2011/OPINION/08/25/ emanuel.weather.predict/index.html?_s=PM:OPINION*.

Merz (2012) discusses uncertainties in flood forecasts. The IPCC (2007) report explains the climate model prediction uncertainties in Figure 4.2. Curry and Webster (2011), Hegerl et al. (2011), and Maslin and Austin (2012) discuss whether the uncertainties are larger.

O'Connor (2008) reviews the history of the human chromosome number. A history of speed-of-light measurements is in Henrion and Fischhoff (1986). The Feynman quote is from his famous "Cargo Cult Science" commencement speech at Caltech in 1974, reprinted in Feynman (1997) and widely available on the web. The video of the gorilla walking through the basketball game is posted at *http://viscog.beckman.illinois.edu/media/ig.html*. The knee surgery study is from Moseley et al. (2002).

General introductions to uncertainty in physical measurements are given by Bevington and Robinson (1992) and Taylor (1997). Entertaining general discussions of problems in predicting the future are given by Gardner (2010) and Silver (2012). The issue of overfitting is discussed by Silver (2012) and *http://blog.lokad.com/journal/2009/4/22/overfitting-when-accuracy-measure -goes-wrong.html* (e.g. the figure in question 5.5).

Hall (2011) describes the L'Aquila events. Murphy (1988) discusses the skill score, and Murphy (1993) and Jollife and Stephenson (2012) discuss weather forecast testing. Morss and Hayden (2010) discuss the "certain death" forecast. Stein et al. (2012) and Stein and Geller (2012) discuss communicating uncertainties in earthquake hazards. Stein et al. (2012) discuss issues in testing earthquake hazard models.

Other relevant papers include Albarello and D'Amico (2008), Miyazawa and Mori (2009), Stirling and Gerstenberger (2010), and Kossobokov and Nekrasova (2012). Projects testing earthquake predictability include Schorlemmer et al. (2010), CSEP (2011), Marzocchi and Zechar (2011) and Kagan and Jackson (2012).

References

Albarello, D., and V. D'Amico, Testing probabilistic seismic hazard estimates by comparison with observations: an example in Italy, *Geophys. J. Int.*, *175*, 1088–1094, 2008.

Bevington, P. R., and D. K. Robinson, *Data Reduction and Error Analysis for the Physical Sciences*, Second Edition, McGraw-Hill, New York, 1992.

CSEP (Collaboratory for the Study of Earthquake Predictability), http://www.cseptesting.org/home, 2011.

Curry, J., and P. Webster, Climate science and the uncertainty monster, *Bull. Am. Meteorol. Soc.*, *92*, 1667–1682, doi: 10.1175/2011BAMS3139.1, December 2011.

Feynman, R. P., *QED: The Strange Theory of Light and Matter*, Princeton University Press, Princeton, NJ, 1986.

Feynman, R. P., *Surely You're Joking, Mr. Feynman! (Adventures of a Curious Character)*, Norton, New York, NY, 1997.

Gardner, D., *Future Babble: Why Expert Predictions Fail – and Why We Believe Them Anyway*, McClelland & Stewart, Toronto, 2010.

Hall, S., Scientists on trial: at fault? *Nature*, *477*, 264–269, 2011.

Hegerl, G., P. Stott, S. Solomon, and F. Zwiers, Comment on "Climate science and the uncertainty monster" by J. Curry, and P. Webster, *Bull. Am. Meteorol. Soc.*, *92*, 1683–1685, doi: 10.1175/2011BAMS3139.1, December 2011.

Henrion, M., and B. Fischhoff, Assessing uncertainty in physical constants, *Am. J. Phys.*, *54*, 791–798, 1986.

Hirschberg, P., E. Abrams, A. Bleistein, W. Bua, L. Monache, T. Dulong, J. Gaynor, B. Glahn, T. Hamill, J. Hansen, D. Hilderbrand, R. Hoffman, B. Morrow, B. Philips, J. Sokich, and N. Stuart, A weather and climate enterprise strategic implementation plan for generating and communicating forecast uncertainty information, *Bull. Am. Meteorol. Soc.*, *92*,1651–1666, 92, 2011.

IPCC, Intergovernmental Panel on Climate Change, *Climate Change 2007: Working Group I: The Physical Science Basis*, Cambridge Univ. Press, Cambridge, 2007.

Jollife, I., and D. Stephenson, *Forecast Verification: A Practioner's Guide in Atmospheric Science*, Wiley, Hoboken, NJ, 2012.

Kagan, Y. Y., and D. D. Jackson, Whole Earth high-resolution earthquake forecasts, *Geophys. J. Int.*, *190*, 677–686, 2012.

Kanamori, H., Prepare for the unexpected, *Nature*, *473*, 147, 2011.

Kossobokov, V. G., and A. K. Nekrasova, Global seismic hazard assessment program maps are erroneous, *Seismic Instrum.*, *48*, 162–170, 2012.

Marzocchi, W., and J. D. Zechar, Earthquake forecasting and earthquake prediction: different approaches for obtaining the best model, *Seismol. Res. Lett.*, *82*, 442–448, doi: 10.1785/gssrl.82.3.442, 2011.

Maslin, M., and P. Austin, Climate models at their limit, *Nature*, *486*, 183–184, 2012.

Merz, B., Role and responsibility of geoscientists for mitigation of natural disasters, *European Geosciences Union General Assembly*, Vienna, 2012.

Miyazawa, M., and J. Mori, Test of seismic hazard map from 500 years of recorded intensity data in Japan, *Bull. Seismol. Soc. Am.*, *99*, 3140–3149, 2009.

Morss, R., and M. Hayden, Storm surge and "certain death": interviews with coastal residents following Hurricane Ike, *Weather, Clim., Soc.*, *2*, 174–189, 2010.

Morss, R., J. Demuth, and J. Lazo, Communicating uncertainty in weather forecasts: a survey of the U.S. public, *Weather Forecast.*, *23*, 974–991, 2008.

Moseley, J. B., K. O'Malley, N. J. Petersen, T. J. Menke, B. A. Brody, D. H. Kuykendall, J. C. Hollingsworth, C. M. Ashton, and N. P. Wray, A controlled trial of arthroscopic surgery for osteoarthritis of the knee, *NEJM*, *347*, 81–88, 2002.

Murphy, A. H., Skill scores based on the mean square error and their relation to the correlation coefficient, *Mon. Weather Rev.*, *116*, 2417–2424, 1988.

Murphy, A. H., What is a good forecast? an essay on the nature of goodness in weather forecasting, *Weather Forecast.*, *8*, 281–293, 1993.

O'Connor, C., Human chromosome number, *Nat. Educ.*, *1*, 1–4, 2008.

Oreskes, N., Why predict? Historical perspectives on prediction in earth science, in *Prediction: Science, Decision Making, and the Future of Nature*, edited by D. Sarewitz, R. Pielke, Jr., and R. Byerly, Jr., pp. 23–40, Island Press, Washington D.C., 2000.

Pilkey, O. H., and L. Pilkey-Jarvis, *Useless Arithmetic: Why Environmental Scientists Can't Predict the Future*, Columbia University Press, New York, 2007.

Powell, C., *It Worked for Me: In Life and Leadership*, Harper, New York, 2012.

Sarewitz, D., R. Pielke, Jr., and R. Byerly, Jr., *Prediction: Science, Decision Making, and the Future of Nature*, Island Press, Washington D.C., 2000.

Schorlemmer, D., J. D. Zechar, M. Werner, D. D. Jackson, E. H. Field, T. H. Jordan, and the RELM Working Group, First results of the Regional Earthquake Likelihood Models Experiment, *Pure Appl. Geophys.*, *167*(8/9), 859–876, doi: 10.1007/s00024-010-0081-5, 2010.

Silver, N., *The Signal and the Noise*, Penguin, New York, 2012.

Stein, S., and R. Geller, Communicating uncertainties in natural hazard forecasts, *Eos Trans. AGU*, *93*, 361–362, 2012.

Stein, S., R. J. Geller, and M. Liu, Why earthquake hazard maps often fail and what to do about it, *Tectonophysics*, *562–563*, 623–626, 2012.

Stirling, A., Keep it complex, *Nature*, *468*, 1029–1031, 2010.

Stirling, M. W., and M. Gerstenberger, Ground motion-based testing of seismic hazard models in New Zealand, *Bull. Seismol. Soc. Am.*, *100*, 1407–1414, 2010.

Taylor, J. R., *An Introduction to Error Analysis: The Study of Uncertainties in Physical Measurements*, University Science Books, Sausalito, CA, 1997.

6

Human Disasters

"The essence of the this-time-it's-different syndrome is simple. It is rooted in the firmly held belief that financial crises are things that happen to other people in other countries at other times; crises do not happen to us, here, and now. We are doing things better, we are smarter, we have learned from past mistakes."

C. Reinhart and K. Rogoff, *This Time Is Different*[1]

6.1 Assessing Hazards

Natural disasters involve humans, in that they occur when a natural process causes loss of life and property. Other disasters are purely human, in that they result from the failure of a human system. Because human disasters have much in common with natural disasters, we can learn a lot about assessing and mitigating natural disasters by examining human disasters.

The two disaster types are similar because many human systems have become so complicated that it is hard for the people in charge to understand fully how they work. As a result, unexpected disasters happen. Assessing the hazard of such disasters has many of the same challenges as assessing those of natural disasters. Similarly, just as societies often prove more vulnerable to natural disasters than expected, human systems often turn out to be much more vulnerable than expected, for analogous reasons.

[1] Reinhart and Rogoff, 2009. Reproduced with permission of Princeton University Press.

Playing against Nature: Integrating Science and Economics to Mitigate Natural Hazards in an Uncertain World, First Edition. Seth Stein and Jerome Stein.
© 2014 John Wiley & Sons, Ltd. Published 2014 by John Wiley & Sons, Ltd.
Companion Website: www.wiley.com/go/stein/nature

An important similarity is the difficulty in assessing disaster probabilities. As we have seen, natural hazard assessments often underestimate their uncertainties. Similarly, those who design and operate human systems often underestimate the risk of failure.

One famous case is the ocean liner *Titanic*, which was considered "unsinkable" (probability of loss zero) because it would float even if four of its sixteen watertight compartments flooded. However, it struck an iceberg on its maiden voyage in April 1912 and sank with the loss of about 1500 lives.

A similar situation arose in the US space shuttle program. In 1986, shuttle *Challenger* exploded on launch on the 51st shuttle flight (Figure 6.1). The explosion resulted from NASA's decision to launch on a cold day, despite knowing that the rubber O-rings between sections of the booster rockets weakened and leaked hot gas in cold temperatures. Physicist Richard Feynman criticized NASA management's assessment that the risk of failure was 1/100,000. As he pointed out in his report dissenting from the government commission, this rate implies that

one could put up a shuttle every day for 300 years expecting to lose only one, we could properly ask what is the cause of management's fantastic faith in the

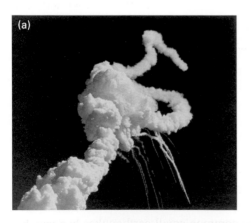

Figure 6.1 (a) Explosion of shuttle *Challenger*. (b) Final launch of shuttle *Columbia*. These two losses in 107 missions correspond to a loss rate of about 1 per 50 missions, much higher than the 1 in 100,000 that had been assumed by NASA management. ((a) NASA, 1986. (b) NASA, 2003.)

machinery . . . One reason may be an attempt to assure the government of NASA perfection and success in order to assure the supply of funds. The other may be that they sincerely believed it to be true, demonstrating an almost incredible lack of communication between themselves and their working engineers.

Feynman argued that the engineers' estimate of the risk, 1000 times greater or 1/100, was more realistic. This assessment proved tragically correct when, in February 2003, shuttle *Columbia* was also lost, on the 107th shuttle mission.

6.2   Vulnerability and Interconnections

Another important similarity between natural and human disasters is that unrecognized vulnerability often arises in similar ways.

Some human disasters arise from single and often embarrassing causes. On June 4, 1996 an unmanned Ariane 5 rocket launched by the European Space Agency exploded 40 seconds after lift-off. The rocket and its cargo were valued at $500 million. The cause of the failure was an error in the software arising from converting a floating point number relating to the horizontal velocity to an integer. As a result, the main and backup computers shut down, so the rocket went out of control and broke apart. Another software error doomed the $125 million Mars Climate Observer mission in 1999, which burned up in the Martian atmosphere. The problem was that software controlling the orbiter's thrusters calculated the force the thrusters needed to exert in pounds, while another piece of software assumed that these values were in the metric unit, newtons. Although resulting discrepancies were noted during the flight, they were ignored.

In many cases, however, human disasters result from interconnections; one failure causes others or several occur simultaneously. A famous analysis, starting from the accident that crippled the Three Mile Island nuclear power plant in 1979, termed such accidents "normal accidents" in that they result from a combination of factors, many of which alone are common and in isolation would not cause disaster. This view is reminiscent of the chaos model for earthquakes, in which one of the common small earthquakes happens by chance to grown into a large one.

Because of this effect, disasters that were thought to be very unlikely occur far more often than would have been expected by treating each portion of a system as independent, such that the probability of all occurring is the product of the probabilities of each occurring alone (section 4.2). This chain-reaction effect is common in technological accidents but also occurs in natural disasters. It is described by Murphy's Law as "anything that can go wrong, will, at the worst time." This law is said to be named after Edward Murphy, one of a team that in 1947 used a rocket-powered sled to learn how much deceleration a pilot could survive. When the measuring instruments in a test read zero because they were installed backwards, Murphy made the observation bearing his name.

Planning for failure chains is hard, because it is often difficult to identify these vulnerabilities in advance, as shown by the sinking of the ocean liner *Andrea Doria* in 1956. Such ships were considered unsinkable – invulnerable – for several reasons. First, they could not collide, because radar let their crews see in night and fog. Moreover, they were divided into watertight compartments and designed to float even if multiple compartments flooded. Still, Murphy's Law prevailed. On a foggy night, as the *Doria* sped toward New York, the Swedish liner *Stockholm* rammed it. Apparently, one or both crews misread poorly designed radar displays. Many things went wrong when the *Stockholm's* strong bow, designed to break through ice, hit the *Doria*. The bulkheads between compartments were designed on the assumption that if the *Doria* took on water, it could not tip, or list, more than 15°. However, these calculations assumed that once fuel tanks were emptied, they would be filled with seawater that provided weight to keep the ship level. In fact, because this required cleaning the tanks in port and disposing of the oily water, it was not done. After the collision, water poured into the tanks on one side, while those on the other stayed empty. Immediately, the ship listed 18°. Water from flooded compartments poured over the tops of the bulkheads and knocked out the generators needed to power the pumps, so flooding continued and the ship listed further. Lifeboat systems were also designed for a list of less than 15°, so boats on the high side could not be launched. Those on the low side were too far from the ship for people to get into, and so were lowered empty and people had to jump down to them. Fortunately, the collision occurred on a calm summer night and other ships used their boats to rescue the passengers and crew.

Similar failure chains often arise in natural disasters, making vulnerability greater than anticipated. The fires after the 1906 San Francisco earthquake, which are thought to have done more damage than the actual shaking, are a prime example. Among the buildings that collapsed was a firehouse, in which the fire chief was killed. Fires broke out as natural gas lines burst and cooking stoves toppled. Firemen connected hoses to hydrants, but no water came out, because all but one of the water mains had broken. Attempts to dynamite firebreaks produced more fires. The fires burned for two days until stopped by a combination of rain, firefighting, and running out of things to burn. When all was over, much of the city was in ruins.

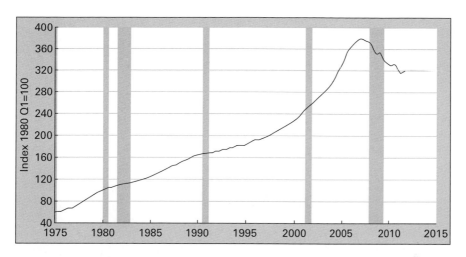

Figure 6.2   US house price index from 1975 to 2011. Prices are nominal, i.e. not adjusted for inflation. Shaded areas indicate US recessions. ("FRED®" charts © Federal Reserve Bank of St. Louis, 2014. All rights reserved. All "FRED®" charts reprinted by permission. http://research.stlouisfed.org/fred2/)

6.3   The 2008 US Financial Disaster

The US financial disaster growing out of the collapse of the housing market in 2006 had major similarities to a natural disaster.

First, the hazard was underestimated. From 1975 to 2007, house prices grew steadily (Figure 6.2), with the rate of growth increasing dramatically from 1995 to 2007. The house price bubble was fueled by low interest rates and subprime loans (i.e. mortgages issued to borrowers who had a high risk of default). Such borrowers were often called "NINJA"s for "no income, no job or assets." Borrowers were encouraged to lie about their finances via "no-doc" or "liar-loan" applications that would not be checked. The share of loans with full documentation decreased from 69% in 2001 to 45% in 2006.

The obvious hazard was that housing prices might collapse. Although it was recognized that many borrowers were unlikely to be able to make loan payments from their income once initial low "teaser" rates ended, it was assumed that the loans would be refinanced from the appreciation of house value. This could only work if housing prices continued to rise in excess of the interest rate. About half of subprime mortgages taken out in 2006 were designed to generate cash for borrowers by refinancing an existing mortgage into a larger mortgage loan. Government policies facilitated these risky loans. Neither Washington nor Wall Street recognized that continued borrowing to refinance without the income to pay the loan was an unsustainable "free lunch." People were making so much money that they ignored economist Herbert Stein's famous dictum "if something can't go on forever, it won't." As in previous "bubbles," including the Dutch tulip bubble of 1637 or the 1995–2001 US "dot-com" stock bubble, prices rose rapidly and then collapsed. As Sir Isaac Newton, who lost heavily in England's 1720 South Sea bubble, admitted, "I can calculate the movement of stars, but not the madness of men."

A second problem was that vulnerability was ignored. The vulnerability was produced by trillions of dollars of risky mortgages that had become embedded throughout the financial system. Slices of mortgage-related securities, called *derivatives* because their values depended on the mortgages, were packaged, repackaged, and sold to investors around the world. Giant financial firms held packages of derivatives either directly or indirectly through investment in hedge funds. The purchases were financed by short-term bank loans. Neither the firms nor the banks worried about the rising debt, because their equity was rising as house prices rose. Charles Prince, the former CEO of Citigroup, told the Financial Crisis Inquiry Commission,

> As more and more subprime mortgages were created as raw material for the securitization process, more and more of it was of lower and lower quality. At the end of that process, the raw material going into it was actually of bad quality, it was toxic quality, and that is what ended up coming out the other end of the pipeline.

These risky derivatives made the entire financial system vulnerable because of an effect called *leverage*. This is a crucial variable in the financial sector; expected growth and vulnerability depend on firms' leverage ratios L, defined as the ratio of their assets, A, to net worth, X:

$$L = A/X = A/(A - D) \qquad (6.1)$$

where net worth is the difference between assets and debts, D. If much of a firm's assets were bought by borrowing money that incurred debt, its net worth is much less than its assets. Hence L is large and the firm is heavily leveraged.

Over time, the fractional change in a firm's net worth depends on the difference between the return on its investments and the interest on its debt, which is the cost of borrowing:

$$\frac{dX}{X} = \frac{d(A - D)}{X} = \frac{dA}{X} - \frac{dD}{X} = \frac{dA}{A}\frac{A}{X} - \frac{dD}{D}\frac{D}{X} = RL - i(L - 1) \qquad (6.2)$$

where $R = dA/A$ is the return on investments due to the productivity of the assets and the capital gain due to the change in their value, and $i = dD/D$ is the interest rate on the debt.

A drop in asset value, shown by a negative R, can be viewed as a financial hazard due to market changes. The resulting drop in net worth is the risk to the firm. Because the change in net worth depends on the change in asset value R times the leverage L, higher leverage makes the firm more vulnerable to a drop in asset value. By analogy to natural hazards:

$$\text{risk} = (\text{hazard})(\text{vulnerability}) = (\text{drop in asset value})(\text{leverage}). \quad (6.3)$$

In 2007 the major investment banks – Bear Stearns, Goldman Sachs, Lehman Brothers, Merrill Lynch and Morgan Stanley – were operating with leverage ratios as high as 40. Thus a 3% drop in a firm's asset values would wipe it out ($-0.03 \times 40 = -1.2$). These high leverage ratios made the economy vulnerable and helped convert the subprime crisis in the housing industry into widespread disaster. When asset values declined, these firms' net worth declined dramatically. Some (e.g., Lehman Brothers) went bankrupt and others (e.g., Bear Sterns) survived only via subsidized purchases or direct government bailouts.

Even in 2006 when housing prices collapsed, the collapse was initially not viewed as having serious consequences. "We think the fundamentals of the expansion going forward still look good," Timothy Geithner, president of the Federal Reserve Bank of New York, told his colleagues in December 2006. This optimism proved incorrect. Increased numbers of foreclosures led to the collapse of securities based on housing prices. By 2008, the financial system was in crisis, as shown in Figure 6.3 by the Financial Stress Index, a composite of financial indices including short- and long-term government and corporate interest rates. A number of major financial institutions were made bankrupt or illiquid, inducing government intervention. Unemployment soared from less than 5% in 2007 to 10% in 2009, and remained above 8% until late 2012.

A third similarity to natural disasters is that the hazard and vulnerability were not recognized, as a result of using erroneous models. Since the 1970s, sophisticated mathematical models were used to develop arcane new financial instruments. Few within the industry beyond their practitioners, termed "quants," understood how the models worked. Nonetheless, as described by Fischer Black, a leader in developing them, the models were "accepted not because it is confirmed by conventional empirical tests, but because researchers persuade one another that the theory is correct and relevant." This acceptance was illustrated by the award in 1997 of the Nobel Prize in economics to Myron Scholes and Robert Merton for work based upon Black's, who had

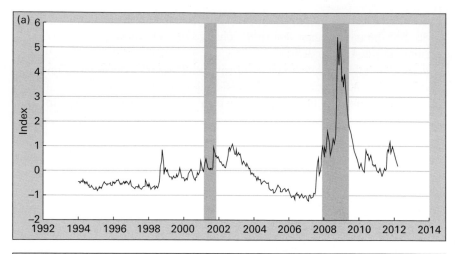

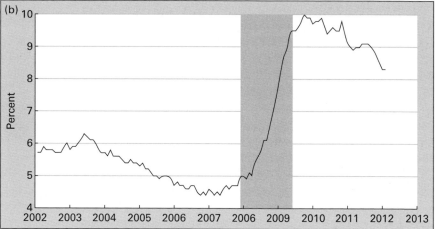

Figure 6.3   Financial Stress Index (a) and unemployment rate (b) showing the effects of the 2008 disaster. Shaded areas indicate US recessions. ("FRED®" charts © Federal Reserve Bank of St. Louis, 2014. All rights reserved. All "FRED®" charts reprinted by permission. http://research.stlouisfed.org/fred2/)

died a few years earlier. Only a year later, Long Term Capital Management, a hedge fund whose directors included Scholes and Merton, collapsed and required a government-organized $3.6 billion bailout. Unfortunately, this collapse did not lead to reassessment of the financial models, whose continued use in developing mortgage backed securities contributed significantly to the 2008 crisis.

These models incorrectly assumed that slicing mortgages and bundling them into derivatives (known as "tranching") made risky assets safe. This "apples in the basket" model assumed that one rotten apple implies nothing about the others. Thus the probability that a few mortgages going bad would cause the entire tranche to fail was assumed to be small. Based on this model, credit rating agencies gave high ratings to risky derivatives. However, the risk of default on mortgages was not independent, because if house prices stopped rising fast enough, or the economy faltered, many borrowers would be unable to pay. Instead, a better model would have been "bread in the loaf "; one moldy slice implies that the next – or the rest of the loaf – is also moldy. Packages of toxic assets, rather than riskless, were actually very risky.

This risk could have been reduced, if proposals to regulate derivatives trading had been adopted. In 1998 the Commodities Futures Trading Commission sought to regulate this trading, and the General Accounting Office concluded that "the sudden failure or withdrawal of any one of these dealers could cause liquidity problems in the markets and could also pose risks to the others, including federally insured banks and the system as a whole." However, Federal Reserve Board chairman Alan Greenspan, Secretary of the Treasury Robert Rubin, and others opposed regulation. Greenspan said ". . . regulation of derivatives transactions that are privately negotiated by professionals is unnecessary . . . By far the most significant event in finance during the past decade has been the extraordinary development and expansion of financial derivatives." As is often the case, the "experts" were proved wrong.

The similarities with the Tohoku earthquake are striking. Considering only too-short records led in one case to the view that house prices could only rise, and in the other to underestimating the size of a possible tsunami. Neglecting interconnections led in one case to the view that mortgage defaults would be uncorrelated, and in the other to the assumption that only single trench segments could fail at one time. In both cases, untested hazard models that had been adopted on faith proved inadequate.

6.4 Pseudodisasters and Groupthink

Communities sometimes experience "pseudodisasters," situations where a disaster is incorrectly viewed as underway or imminent. For our purposes, these can be viewed as situations where a hazard is greatly overassessed and then overmitigated, often at considerable cost. Two classic cases illustrate the pattern.

In 1976, some scientists and physicians at the US government's Centers for Disease Control (CDC) warned of an upcoming swine flu epidemic. Although many of the staff thought an epidemic was unlikely, their views were ignored. In the *Washington Post*'s words, "the rhetoric of risk suffered steady inflation as the topic moved from the mouths of scientists to the mouths of government officials." Soon the head of the CDC discussed the "strong possibility" of an epidemic. Staff in the Department of Health, Education and Welfare (HEW) announced that "the chances seem to be 1 in 2" and the secretary of HEW told the White House that "projections are that this virus will kill one million Americans in 1976."

Some scientists, including Albert Sabin who developed the oral polio vaccine, called for stockpiling the vaccine but only using it if an epidemic occurred. These calm voices had no effect. Perhaps due to the upcoming election, President Gerald Ford's administration launched a program to vaccinate the entire population (Figure 6.4). At a cost of millions of dollars, 40 million Americans were vaccinated before the program was suspended due to reactions to the vaccine. About 500 people had serious reactions and 25 people died; by comparison, one person died from swine flu. Ford lost the election, so the only immediate winners were the companies that made the vaccine.

The fact that the dreaded "aporkalypse" did not materialize helped the public health system to respond somewhat more thoughtfully to new exotic strains such as the 2003 bird flu and 2009 swine flu. However, warnings of doom still recur regularly. In 2005, officials warned of an upcoming bird flu pandemic that could kill 150–200 million people worldwide, with one giving the "best case scenario" as 7.4 million deaths. Some Americans stockpiled

Figure 6.4   President Gerald Ford is vaccinated against swine flu in 1976. (Kennerly, 1976. Courtesy of Gerald R. Ford Library.)

food, water, medicine, and guns. A book entitled *The Bird Flu Preparedness Planner* warned, "Bird Flu is real. It's deadly. And it's spreading. Within the next year it could threaten your home, your livelihood, your family and even your life." Someone stockpiling months of food explained that "there will be gangs just looting. . . so there will be four male adults in this house who know how to use firearms." Sales of Tamiflu, an antiviral medicine, soared. As usual, the pandemic did not occur.

In the late 1990s, some computer experts warned of an impending catastrophe on January 1, 2000. Supposedly computer systems would fail, because they represented dates in their memory using only two digits. Hence the year 2000 ("Y2K") would not be distinguished from 1900. Books and the media warned of the upcoming disaster due to the collapse of the financial system, public utilities, and other institutions of technological society (Figure 6.5). Many warned that Y2K would cause TEOTWAWKI – "the end of the world as we know it."

The hype grew despite some calm voices that included Microsoft's Bill Gates, who predicted no serious problems and blamed "those who love to tell tales of fear" for the scare. *Newsweek* carried a cover story titled "The Day the World Shut Down: Can We Fix the 2000 Computer Bug Before It's Too Late?" *Vanity Fair* magazine claimed that "folly, greed and denial" had "muffled two decades of warnings from technology experts." Broadcast minister Jerry Falwell sold a $28 video titled "A Christian's Guide to the Millennium Bug" and warned that "Y2K may be God's instrument to shake this nation, to humble this nation." Television programs, including some starring *Star Trek's* Leonard Nimoy, hyped the impeding catastrophe. Survivalists stockpiled food, water, and guns. Y2K insurance policies were written.

Senator Daniel Moynihan wrote to President Clinton urging that the military should take command of dealing with the problem. The US government established the President's Council On Year 2000 Conversion, headed by the "Y2K czar." Special legislation was passed and government agencies led by the Federal Emergency Management Agency (FEMA) swung into action. FEMA, in turn, mobilized state emergency management agencies.

On the dreaded day, the "millennium bug" proved a bust. Nations and businesses that had prepared aggressively watched sheepishly as few – and only minor – problems occurred among the many countries and businesses that had taken few or no precautions. However, the estimated $300 billion spent made Y2K preparations very profitable for the computer industry. Y2K programming work shipped from the US fueled the growth of India's profitable outsourcing industry at the expense of US programmers' jobs. On the plus side, there was a long-term benefit in that claims of impending computer-related disasters are now treated more calmly.

Figure 6.5   One of many books advising how to survive the predicted Y2K disaster.

The swine flu and Y2K pseudodisasters illustrate a common problem in policy making, a consensus approach called "groupthink" that often causes errors. Political scientists' and management experts' studies find that in groupthink, members of a group start with absolute belief that they are right. They talk only to each other, convince each other using specious arguments, pressure others to agree, refuse to consider clear evidence that contradicts their views, and ignore outside advice. They more they talk to each other, the more convinced they become. Well-studied examples include the US government's decisions to launch the failed invasion of Cuba at the Bay of Pigs, go to war unsuccessfully in Vietnam and Iraq, and launch shuttle *Challenger* in conditions known to be dangerous.

The hazard assessments for Fukushima (section 2.3) and the financial crisis (section 6.3) also involved groupthink. Those in charge talked themselves into underestimating or ignoring the many uncertainties involved in choosing their policies.

6.5   Disaster Chic

The swine flu and Y2K pseudodisasters also illustrate the problem of apocalyptic predictions. Some researchers think that unneeded public fears are becoming more common. Sociologist Barry Glassner, author of *The Culture of Fear*, argues that "Crime rates plunge, while surveys show Americans think that crime rates are rising. At a time when people are living longer and healthier, people are worried about iffy illnesses."

Researchers feel that such fears are becoming more serious for several reasons. First, people and groups with agendas feed them. In *False Alarm: The truth about the epidemic of fear*, Marc Siegel from New York University's medical school explains, "Bureaucrats use fear by playing Chicken Little. They can claim at every turn that a disaster may be coming, thereby keeping their budgets high and avoiding blame should disaster really strike." Cass Sunstein of the University of Chicago law school points out in *Risk and Reason* that an official "may make speeches and promote policies that convey deep concern about the very waste spill that he actually considers harmless."

A famous example is the fear of child abductions. Since the 1980s parents have been warned in the media that their children are at high risk of being abducted. A National Center for Missing and Exploited Children was set up, anti-abduction groups formed, and special legislation was passed. Police departments fingerprinted children to aid in their recovery if they were kidnapped. Entrepreneurs offered products to address the problem, including courses to educate children about "Stranger Danger" and special watches with an alert signal and flashing light that children could use to call attention if they were being kidnapped. Those promoting solutions to the problem somehow neglected to point out that of about 75 million children in the US, only about 100 a year are abducted by strangers, so the risk is quite small.

Problems often arise with large government contracts set up to deliver protection from hypothesized threats. Often those promoting the alleged threat will gain from the expensive solution. The Los Angeles Times described how, for example, Richard Danzig, a former Secretary of the Navy who consulted for the Pentagon and the Department of Homeland Security, urged the government to buy a drug that would help to defend the US from a biological attack using a form of anthrax designed to resist antibiotics, although no such bacterium is known to exist. While doing so, he was paid more than $1 million as a director of the biotechnology firm that won $334 million in contracts to supply the drug.

Second, the media play up fears. In most cases they know the story is hype but the dynamics of their business encourages them not to say so. There is an incentive not to ask hard questions. As a journalist explains in *False Alarm*, "I don't like reporting the overhyped stories that unnecessarily scare people, but these are my assignments." Seigel's assessment is, "Many news teasers use the line 'Are you and your family at risk?' The answer is usually no, but that tagline generates concern in every viewer and keeps us tuned in. If we didn't fundamentally misunderstand the risk, we probably wouldn't watch."

One of many examples is the way the news media declared the summer of 2001 as "the summer of the shark." As reporter John Stossel explained, "Instead of putting risks in proportion, we [reporters] hype interesting ones. Tom Brokaw, Katie Couric, and countless others called 2001 the 'summer of the shark.' In truth, there wasn't a remarkable surge in shark attacks in 2001. There were about as many in 1995 and 2000, but 1995 was the year of the O.J.Simpson trial, and 2000 was an election year. The summer of 2001 was a little dull, so reporters focused on sharks."

Third, people like disaster stories, presumably because we enjoy being a little scared. We like disaster movies, Halloween, and roller coasters. People know that these stories are often hype, so we are not surprised when the predicted disaster of the week turns out much smaller than predicted or does not happen at all.

The problem for natural hazard planning is how to decide which of many possible threats to take seriously. We need to combine careful analysis and common sense to try to identify the threats that are most serious from a huge number of possible ones, and best use our limited resources to prepare for them. As experience shows, deciding how to do this is not easy.

Questions

6.1.   Although the space shuttle was a new program, suggest ways in which the experience of the Apollo program or other analogous experiences could have been used for more realistic risk assessment.

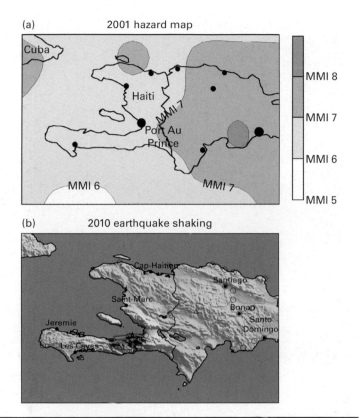

(a) 2001 hazard map

(b) 2010 earthquake shaking

PERCEIVED SHAKING	Not felt	Weak	Light	Moderate	Strong	Very strong	Severe	Violent	Extreme
POTENTIAL DAMAGE	none	none	none	Very light	Light	Moderate	Moderate/Heavy	Heavy	Very Heavy
PEAK ACC.(%g)	<0.17	0.17–1.4	1.4–3.9	3.9–9.2	9.2–18	18–34	34–65	65-124	>124
PEAK VEL.(cm/s)	<0.1	0.1–1.1	1.1–3.4	3.4–8.1	8.1–16	16–31	31–60	60–116	>116
INSTRUMENTAL INTENSITY	I	II-III	IV	V	VI	VII	VIII	IX	X

Figure 1.4 (a) Seismic hazard map for Haiti produced prior to the 2010 earthquake showing maximum shaking expected to have a 10% chance of being exceeded once in 50 years, or on average once about every 500 years. (b) Map of the shaking in the 2010 earthquake. (Stein et al., 2012. Reproduced with permission of Elsevier B.V.)

Playing against Nature: Integrating Science and Economics to Mitigate Natural Hazards in an Uncertain World, First Edition. Seth Stein and Jerome Stein.
© 2014 John Wiley & Sons, Ltd. Published 2014 by John Wiley & Sons, Ltd.
Companion Website: www.wiley.com/go/stein/nature

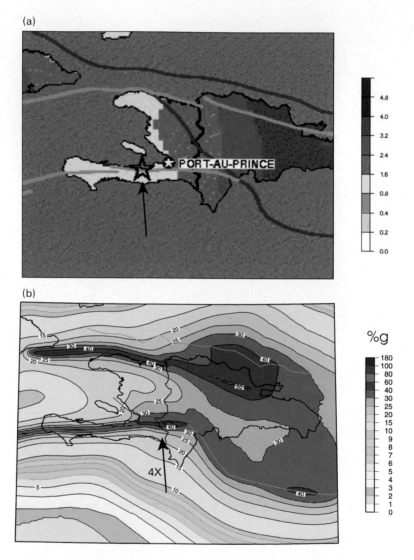

Figure 1.5 Comparison of seismic hazard maps for Haiti made before (a) and shortly after (b) the 2010 earthquake. The newer map shows a factor of four higher hazard on the fault that had recently broken in the earthquake. (Stein et al., 2012. Reproduced with permission of Elsevier B.V.)

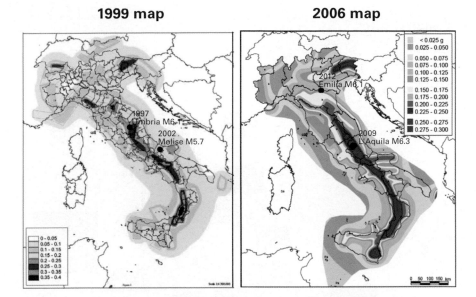

Figure 1.6 Comparison of successive Italian hazard maps, which forecast some earthquake locations well and others poorly. The 1999 map was updated after the missed 2002 Molise earthquake and the 2006 map will presumably be updated because it missed the 2012 Emilia earthquake. (Stein et al., 2013. Reproduced with permission of Elsevier B.V.)

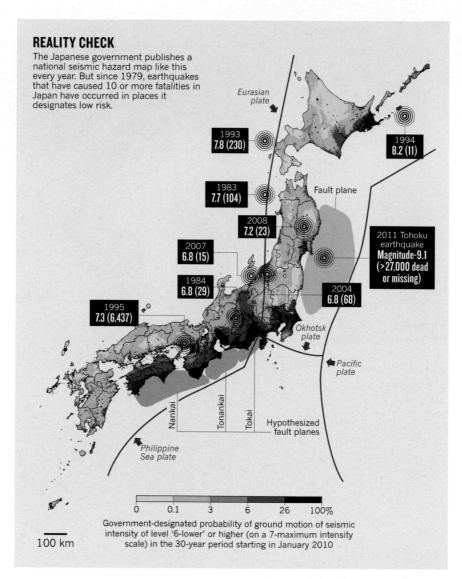

Figure 2.1 Comparison of Japanese government hazard map to the locations of earthquakes since 1979 that caused 10 or more fatalities. Hazard is shown as probability that the maximum ground acceleration (shaking) in any area would exceed a particular value during the next 30 years. Larger expected shaking corresponds to higher predicted hazard. The Tohoku area is shown as having significantly lower hazard than other parts of Japan, notably areas to the south. Since 1979, earthquakes that caused 10 or more fatalities occurred in places assigned a relatively low hazard. (Geller, 2011. Reproduced with permission of *Nature*.)

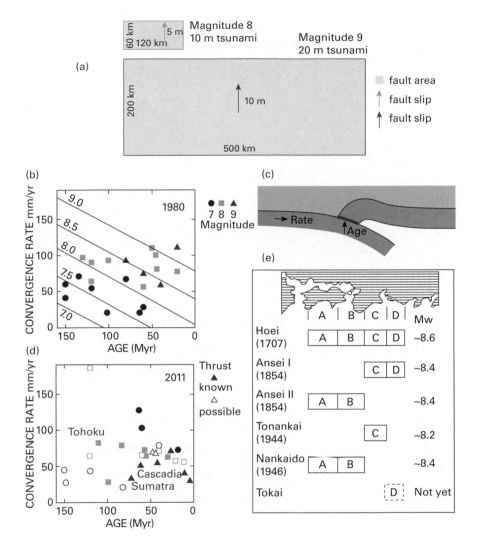

Figure 2.4 What went wrong at Tohoku. (a) Illustration of the relative fault dimensions, average fault slip, and average tsunami run-up for magnitude 8 and 9 earthquakes. (b) Data available in 1980, showing the largest earthquake known at various subduction zones. Magnitude 9 earthquakes had been observed only where young lithosphere subducts rapidly. Diagonal lines show predicted maximum earthquake magnitude. (c) Physical interpretation of this result in terms of strong mechanical coupling and thus large earthquakes at the trench interface. (d) Update of (b) with data including the 2004 Sumatra and 2011 Tohoku earthquakes. (e) Earthquake history for the Nankai trough area illustrating how different segments rupturing cause earthquakes of different magnitudes. Segment "D" is the presumed Tokai seismic gap. (Stein and Okal, 2011. Reproduced with permission of American Geophysical Union.)

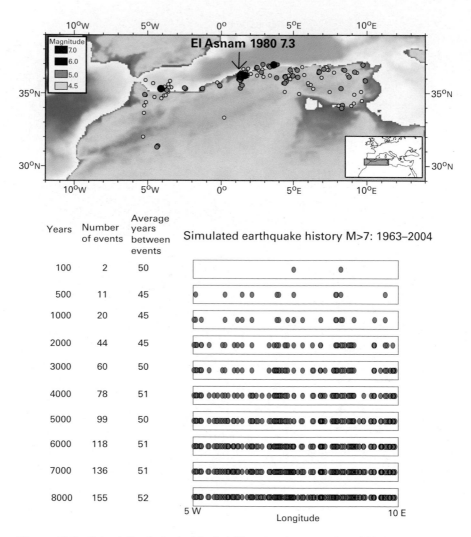

Figure 10.3 Seismicity along the North Africa plate boundary for 1963–2004. Simulations using a frequency-magnitude relation derived from these data predict that if seismicity is uniform in the zone, about 8,000 years of records is needed to avoid apparent concentrations and gaps. (Swafford and Stein, 2007. Reproduced with permission of Geological Society of America.)

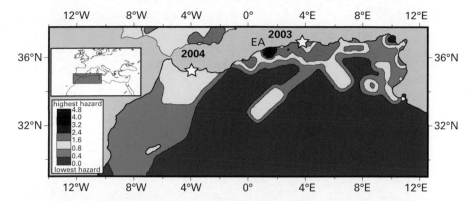

Figure 10.4 Global Seismic Hazard Map (1999) for North Africa, showing peak ground acceleration in m/s² expected at 10% probability in 50 years. Note "bull's-eye" at site of the 1980 M_s 7.3 El Asnam (EA) earthquake. The largest subsequent earthquakes to date, the May 2003 M_s 6.8 Algeria and February 2004 M_s 6.4 Morocco events (stars) did not occur in the predicted high hazard regions. (Swafford and Stein, 2007. Reproduced with permission of Geological Society of America.)

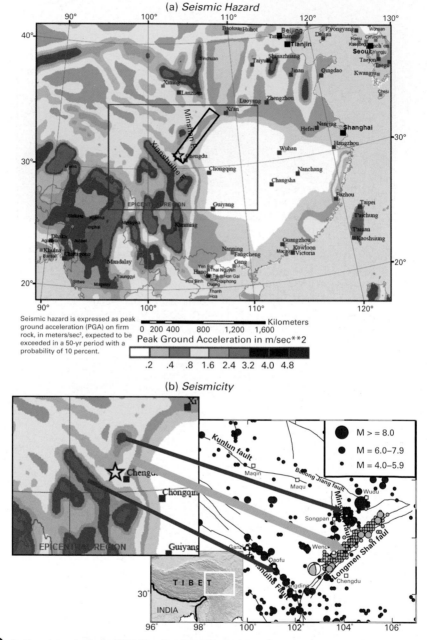

(a) *Seismic Hazard*

Seismic hazard is expressed as peak ground acceleration (PGA) on firm rock, in meters/sec², expected to be exceeded in a 50-yr period with a probability of 10 percent.

Peak Ground Acceleration in m/sec**2

.2   .4   .8   1.6   2.4   3.2   4.0   4.8

(b) *Seismicity*

M >= 8.0
M = 6.0–7.9
M = 4.0–5.9

● Earthquakes prior to the 2008 Wenchuan event
◐ Aftershocks of the Wenchuan event delineating the rupture zone

Figure 10.5   (a) Seismic hazard map for China produced prior to the 2008 Wenchuan earthquake, which occurred on the Longmenshan Fault (black rectangle). (b) Seismicity in the region. The hazard map showed low hazard on the Longmenshan fault, on which little instrumentally recorded seismicity had occurred before the Wenchuan earthquake, and higher hazard on faults nearby that showed more seismicity. (Stein et al., 2012. Reproduced with permission of Elsevier, B.V.)

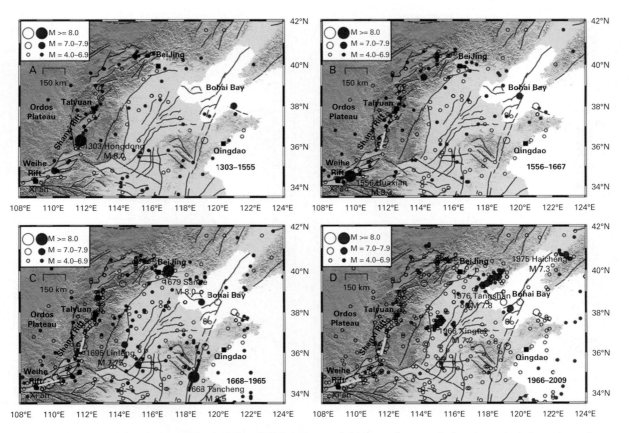

Figure 10.7 Earthquake history of North China. Solid circles are locations of events during the period shown in each panel; open circles are locations of events from 780 BC to the end of the previous period (AD 1303 for panel A). Bars show the rupture lengths for selected large events. (Liu et al., 2011. Reproduced with permission of the Geological Society of America.)

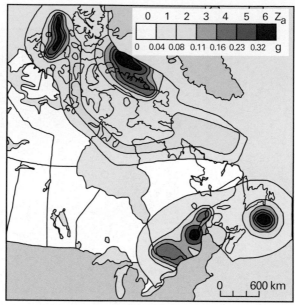

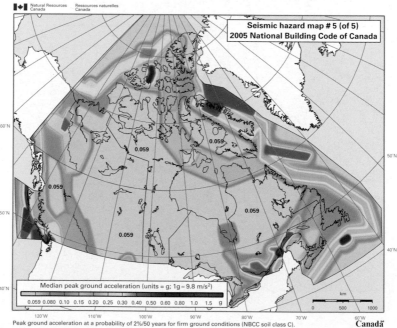

Figure 10.8 Comparison of the 1985 and 2005 Geological Survey of Canada earthquake hazard maps of Canada. The older map shows concentrated high hazard bull's-eyes along the east coast at the sites of the 1929 Grand Banks and 1933 Baffin Bay earthquakes, whereas the new map assumes that similar earthquakes can occur anywhere along the margin. (Stein et al., 2012. Reproduced with permission of Elsevier, B.V.)

(a) Conditional probability of earthquake in next t years

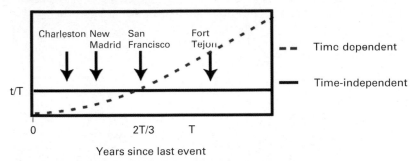

(b) Conditional probabilities of earthquake in next 50 years

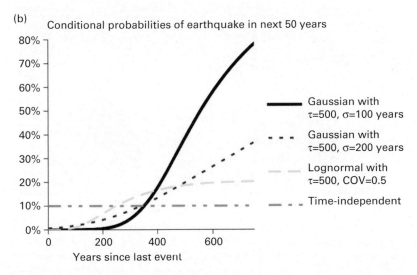

Figure 10.9 Top: Schematic comparison of time-independent and time-dependent models for different seismic zones. Charleston and New Madrid are "early" in their cycles, so time-dependent models predict lower hazards. The two model types predict essentially the same hazard for a recurrence of the 1906 San Francisco earthquake, and time-dependent models predict higher hazard for the nominally "overdue" recurrence of the 1857 Fort Tejon earthquake. The time-dependent curve is schematic because its shape depends on the probability distribution and its parameters. Bottom: Comparison of the conditional probability of a large earthquake in the New Madrid zone in the next 50 years, assuming that the mean recurrence time is 500 years. In the time-independent model the probability is 10%. Time-dependent models predict lower probabilities of a large earthquake for the next hundred years. (Hebden and Stein, 2009. Reproduced with permission of the Seismological Society of America.)

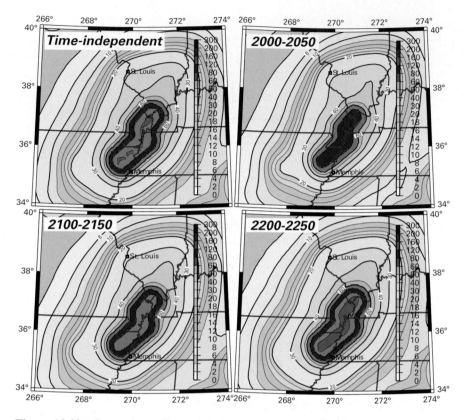

Figure 10.10 Comparison of hazard maps for the New Madrid zone. Shading shows peak ground acceleration as percentages of 1 g. Compared to the hazard predicted by the time-independent model, the time-dependent model predicts noticeably lower hazard for the periods 2000–2050 and 2100–2150, but higher hazard if a large earthquake has not occurred by 2200. (Stein et al., 2012. Reproduced with permission of Elsevier, B.V.)

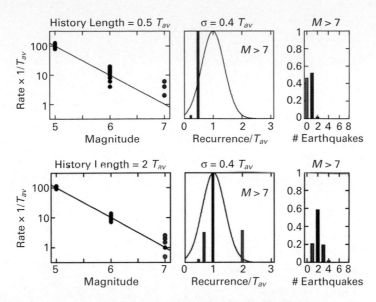

Figure 10.11 Results of numerical simulations of earthquake sequences. Rows show results for sequences of different lengths. Left panels show the log-linear frequency-magnitude relation sampled, with dots showing the resulting mean recurrence times. Center panels show the parent distribution of recurrence times for M ⩾ 7 earthquakes (smooth curve) and the observed mean recurrence times (bars). Right panels show the fraction of sequences in which a given number of M ⩾ 7 earthquakes occurred. (Stein and Newman, 2004. Reproduced with permission of Seismological Society of America.)

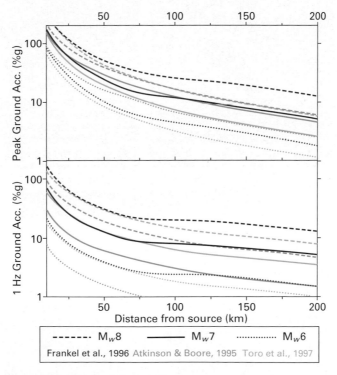

Figure 10.12 Comparison of ground motion (peak ground acceleration and 1 Hz) as a function of distance for different earthquake magnitudes predicted by three models for the central US. For M_w 8, the Frankel et al. (1996) model predicts significantly higher values than the others. (Newman et al., 2001. Reproduced with permission of Seismological Society of America.)

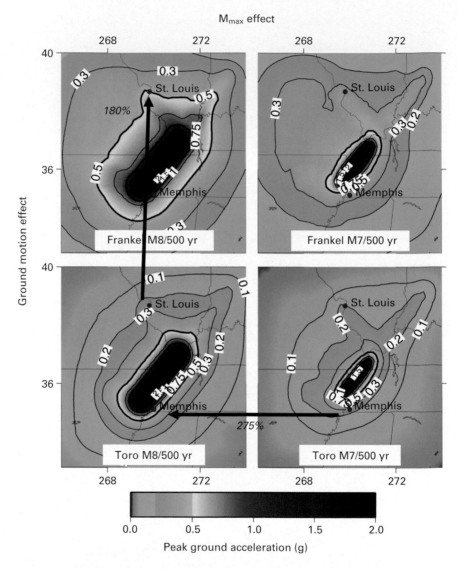

Figure 10.13 Comparison of the predicted hazard (2% probability in 50 years) showing the effect of different ground motion models and maximum magnitudes of the New Madrid fault source. (Newman et al., 2001. Reproduced with permission of Seismological Society of America.)

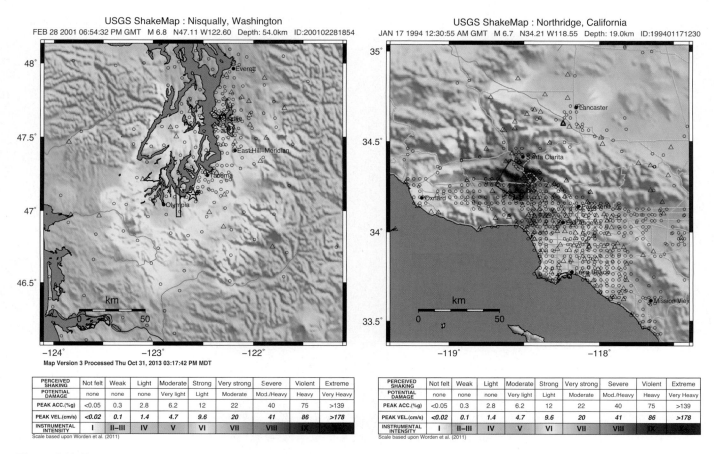

USGS ShakeMap : Nisqually, Washington
FEB 28 2001 06:54:32 PM GMT   M 6.8   N47.11 W122.60   Depth: 54.0km   ID:200102281854

Map Version 3 Processed Thu Oct 31, 2013 03:17:42 PM MDT

PERCEIVED SHAKING	Not felt	Weak	Light	Moderate	Strong	Very strong	Severe	Violent	Extreme
POTENTIAL DAMAGE	none	none	none	Very light	Light	Moderate	Mod./Heavy	Heavy	Very Heavy
PEAK ACC.(%g)	<0.05	0.3	2.8	6.2	12	22	40	75	>139
PEAK VEL.(cm/s)	<0.02	0.1	1.4	4.7	9.6	20	41	86	>178
INSTRUMENTAL INTENSITY	I	II–III	IV	V	VI	VII	VIII	IX	X+

Scale based upon Worden et al. (2011)

USGS ShakeMap : Northridge, California
JAN 17 1994 12:30:55 AM GMT   M 6.7   N34.21 W118.55   Depth: 19.0km   ID:199401171230

PERCEIVED SHAKING	Not felt	Weak	Light	Moderate	Strong	Very strong	Severe	Violent	Extreme
POTENTIAL DAMAGE	none	none	none	Very light	Light	Moderate	Mod./Heavy	Heavy	Very Heavy
PEAK ACC.(%g)	<0.05	0.3	2.8	6.2	12	22	40	75	>139
PEAK VEL.(cm/s)	<0.02	0.1	1.4	4.7	9.6	20	41	86	>178
INSTRUMENTAL INTENSITY	I	II–III	IV	V	VI	VII	VIII	IX	X+

Scale based upon Worden et al. (2011)

Figure Q11.10

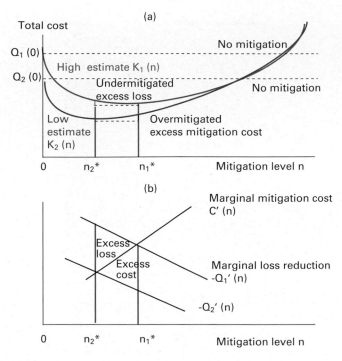

Figure 12.3 Selecting mitigation levels. Panel a): Comparison of total cost curves for two estimated hazard levels. For each, the optimal mitigation level, $n*$, minimizes the total cost, the sum of expected loss and mitigation cost. Panel b): In terms of derivatives, $n*$ occurs when the reduced loss $-Q'(n)$ equals the incremental mitigation cost $C'(n)$. If the hazard is assumed to be described by one curve but is actually described by the other, the assumed optimal mitigation level causes nonoptimal mitigation, and thus excess expected loss or excess mitigation cost. (Stein and Stein, 2013b. Reproduced with permission of the Geological Society of America.)

6.2.   What risk assessment lessons can be learned from the *Andrea Doria's* sinking?

6.3.   What data, starting in 1975, might have been used to test the assumption that US house prices would continue growing as they did?

6.4.   Long before sophisticated but untested financial models were accepted because of consensus – "experts" persuading each other that the theory was correct – Richard Feynman (1965) described testing theories about the physical world (which he called "guesses") this way:

> . . . we compute the consequences of the guess to see what would be implied if this law that we guessed is right. Then we compare the result of the computation to nature, with experiment or experience, compare it directly with observation, to see if it works. If it disagrees with experiment it is wrong. In that simple statement is the key to science. It does not make any difference how beautiful your guess is. It does not make any difference how smart you are, who made the guess, or what his name is – if it disagrees with experiment it is wrong. That is all there is to it.

How might this approach have been applied to financial models?

6.5.   Derive the final result showing the effect of leverage in equation 6.2.

6.6.   Although the pandemics discussed in this chapter did not materialize, pandemics have occurred in the past, including the 1918–1919 influenza pandemic that is estimated to have killed at least 20 million people worldwide. What lessons could the 1976 "aporkalypse" have for future situations when a small outbreak somewhere suggests that a global pandemic may occur?

6.7.   Various explanations have been suggested for why the Y2K issue was so overblown. One is the power of the millennium idea, leading to a view that something important should happen. Another is that people are drawn to apocalyptic visions, which keep resurfacing despite the fact that they do not materialize. A third is that although today many people have had dates on their computers wrong with no ill effects, in 1999 fewer people were familiar with computers. A fourth is that, between the end of the US–USSR Cold War in 1991 and the September 11, 2001 attacks, the formidable US national security bureaucracy was seeking new missions. Which of these seem plausible to you? Are there other explanations that also seem plausible? How would you allocate blame – totaling 100% – between these explanations?

6.8.   Suggest some ways to evaluate whether a threatened disaster, like swine flu, bird flu, or Y2K, should be treated as a serious threat. How should government officials describe the situation to the public?

6.9.   In February 2003, US government officials urged the public to stock up on duct tape and plastic sheeting to seal homes against a biological terrorism attack that never materialized. The public responded, emptying hardware store shelves. How would you respond to a similar warning today and why?

6.10.   In January 2011, the US Department of Homeland Security's Federal Emergency Management Agency (FEMA) issued a call to vendors to "identify sources of supply for meals in support of disaster relief efforts based on a catastrophic disaster event within the New Madrid Fault System for a survivor population of 7M to be utilized for the sustainment of life during a 10-day period of operations." These 14 million meals per day "must have 36 months of remaining shelf life upon delivery." This request, which was canceled promptly once the news media sought information, was never explained. Presumably it had something to do with the government's program to use the anniversary of the large 1811–1812 earthquakes in the area to promote their view that parts of the central US faced earthquake hazards as high as California's, and that buildings should be built to the same costly safety standards. Estimate how much FEMA was planning to spend. What questions would you have asked if you were in the news media?

Further Reading and Sources

Figure 6.1 is from *http://grin.hq.nasa.gov/ABSTRACTS/GPN-2003-00080.html* and *http://grin.hq.nasa.gov/ABSTRACTS/GPN-2004-00012.html*. Feynman's activities on the Challenger commission are described by Gleick (1992) and his dissent to the commission's report is reprinted in Feynman (1988) and available at *http://science.ksc.nasa.gov/shuttle/missions/51-l/docs/rogers-commission/Appendix-F.txt*.

Perrow (1984) introduced the concept of "normal accidents." Chain reaction effects are discussed for technological accidents by Chiles (2002) and for natural disasters by Lomnitz and Castanos (2007).

The sinking of the *Andrea Doria* is described by Moscow (1981). Winchester (2005) describes the aftermath of the 1906 San Francisco earthquake.

Stein (2012) and Stein and Stein (2012) discuss causes of the US financial disaster. Figures 6.2 and 6.3 are from Federal Reserve Bank of St Louis Economic Data *http://research.stlouisfed.org/fred2/*. FCIC (2011) is the Financial Crisis Inquiry Commission report. Geithner's "look good" view is from Applebaum (2012). Derman (2004), Overbye (2009), Salmon (2009), and Stein (2012) discuss the quants' new financial instruments. The Black quotation is from Derman (2004). Long Term Capital Management's collapse is described in Lowenstein (2000).

The pseudodisaster and disaster chic discussions are from Stein (2010). Quotations about swine flu are from Brown (2002, 2005). Warnings about the 2005 bird flu scare are quoted in Sturcke (2005), and Manning (2005) describes people storing food, water, and weapons. The bird flu book mentioned (Woodson, 2005) is still available. The *Newsweek* Y2K cover story ran on June 2, 1997. Senator Moynihan's letter is in the Congressional Record (August 11, 1996) and available online. Falwell's warning is described in the January 14, 1999 issue of *The Economist*. Many of the dire warnings and plans can still be found online by searching "FEMA Y2K" and the like. Dutton (2010) summarizes how the world as we know it didn't end. *The Economist* (December 20, 2012) summarizes end-of-the-world predictions in the past two millennia.

Janis (1982) explores the groupthink phenomenon. The anthrax story is from the *Los Angeles Times* (Willman, 2013). CNN reviewed the "shark summer" story on March 14, 2003 in "Shark summer' bred fear, not facts" at *http://www.cnn.com/2003/TECH/science/03/13/shark.study/*. The duct tape scare is described in the *Christian Science Monitor* (Feldman, 2003), *New York Times* (Easterbrook, 2003), and at *http://www.cnn.com/2003/US/02/11/emergency.supplies/*.

FEMA's call for packaged meals (HSFEHQ-11-R-Meals) is at *https://www.fbo.gov/index?s=opportunity&mode=form&id=40c5ec4c02287bcf1b3a974770d871a1&tab=core&_cview=1*.

References

Appelbaum, B., Inside the Fed in 2006: a coming crisis, and banter, *New York Times*, January 12, 2012.

Brown, D., A shot in the dark: swine flu vaccine's lessons, *Washington Post*, May 27, 2002.

Brown, D., Run on drug for avian flu has physicians worried, *Washington Post*, October 25, 2005.

Chiles, J., *Inviting Disaster: Lessons from the Edge of Catastrophe*, Harper Business, New York, 2002.

Derman, E., *My Life As A Quant: Reflections on Physics and Finance*, Wiley, New York, 2004.

Dutton, D., It's always the end of the world as we know it, *New York Times*, January 1, 2010.

Easterbrook, G., The smart way to be scared, *New York Times*, February 16, 2003.

FCIC, *Financial Crisis Inquiry Report*, Public Affairs, New York, 2011.

Feldman, L., Terror alerts create a run on duct tape, *Christian Science Monitor*, February 13, 2003.

Feynman, R.P., *The Character of Physical Law*. MIT Press, Cambridge, 1965. (online as the final lecture in http://www.openculture.com/2012/08/the_character_of _physical_law_richard_feynmans_legendary_lecture_series_at_cornell_1964 .html)

Feynman, R. P., *What Do You Care What Other People Think*, W. W. Norton, New York, 1988.

Glassner, B., *The Culture of Fear: Why Americans are Afraid of the Wrong Things*, Basic Books, New York, 2000.

Gleick, J., *Genius: The Life and Science of Richard Feynman*, Pantheon, New York, 1992.

Janis, I., *Groupthink: Psychological Studies of Policy Decisions and Fiascoes*, Cengage Learning, Stamford, CT, 1982.

Lomnitz, C., and H. Castanos, Disasters and maximum energy production, in *Continental Intraplate Earthquakes, Science, Hazard, and Policy Issues*, Special Paper 425, edited by S. Stein and S. Mazzotti, pp. 387–396, Geol. Soc. Amer., Boulder, CO, 2007.

Lowenstein, R., *When Genius Failed: The Rise and Fall of Long-Term Capital Management*, Random House, New York, 2000.

Manning, A., Civilians prepare for long-haul pandemic, *USA Today*, December 7, 2005.

Moscow, A., *Collision Course*, Putnam, New York, 1981.

Overbye, D., They tried to outsmart Wall Street, *New York Times*, March 10, 2009.

Perrow, C., *Normal Accidents: Living with High-Risk Technologies*, Basic Books, New York, 1984.

Reinhart, C., and K. Rogoff, *This Time is Different: Eight Centuries of Financial Folly*, Princeton University Press, Princeton, NJ, 2011.

Salmon, F., Recipe for disaster: the formula that killed Wall Street, *Wired Magazine*, February 23, 17.03, 2009.

Siegel, M., *False Alarm: The Truth About the Epidemic of Fear*, Wiley, Hoboken, NJ, 2005.

Stein, J. L., *Stochastic Optimal Control and the U.S. Financial Debt Crisis*, Springer, New York, 2012.

Stein, J. L., and S. Stein, Gray Swans: comparison of natural and financial hazard assessment and mitigation, *Nat. Hazard.*, doi: 10.1007/s11069-012-0388-x, 2012.

Stein, S., *Disaster Deferred: How New Science is Changing our View of Earthquake Hazards in the Midwest*, Columbia University Press, New York, 2010.

Stossel, J., *Give Me a Break*, Harper, New York, 2004.

Sturcke, J., Bird flu pandemic 'could kill 150 m', *The Guardian*, September 30, 2005.

Sunstein, C., *Risk and Reason*, Cambridge University Press, Cambridge, 2002.

Willman, D., Anthrax drug brings $334 million to Pentagon advisor's biotech firm, *Los Angeles Times*, May 19, 2013.

Winchester, S., *A Crack in the Edge of the World: America and the Great California Earthquake of 1906*, Harper Collins, New York, 2005.

Woodson, G., *The Bird Flu Preparedness Planner: What it is. How it spreads. What you can do*, HCI, Deerfield Beach, FL, 2005.

7

How Much Is Enough?

"Systems analysis is a reasoned approach to highly complicated problems of choice in a context characterized by much uncertainty; it provides a way to deal with different values and judgments; it looks for alternative ways of doing a job; and it seeks, by estimating in quantitative terms where possible, to identify the most effective alternative . . . It is not physics, engineering, mathematics, economics, political science, statistics; yet it involves elements of all these disciplines. It is much more a frame of mind."

Enthoven and Smith (1971)[1]

7.1 Rational Policy Making

Making natural hazards policy involves complicated choices intended to mitigate the effects of hazards that are often hard to assess. Ideally, this would be done by carefully assessing the hazard, including the uncertainties involved, formulating a set of possible mitigation options, estimating the costs and benefits of each, and presenting the results to communities so they can make informed choices. In reality, this rarely occurs. Instead, policies are often made based on inadequate assessments of the hazard and without careful consideration of mitigation strategies.

These difficulties are illustrated by a report from the American Society of Civil Engineers assessing the failure of New Orleans' hurricane protection

[1]Enthoven and Smith, 1971.

Playing against Nature: Integrating Science and Economics to Mitigate Natural Hazards in an Uncertain World, First Edition. Seth Stein and Jerome Stein.
© 2014 John Wiley & Sons, Ltd. Published 2014 by John Wiley & Sons, Ltd.
Companion Website: www.wiley.com/go/stein/nature

system as a result of Hurricane Katrina in August 2005. Among the report's conclusions were:

- A large portion of the destruction was caused not only by the storm itself, but by the storm's exposure of engineering and engineering-related policy failures. The levees and floodwalls breached because of a combination of unfortunate choices and decisions, made over many years, at almost all levels of responsibility.
- No single agency was in charge of hurricane protection in New Orleans. Rather, responsibility for the maintenance and operation of the levees and pump stations was spread over many federal, state, parish, and local agencies. This lack of inter-agency coordination led to many adverse consequences.
- The hurricane protection system was constructed as individual pieces – not as an interconnected system – with strong portions built adjacent to weak portions, some pump stations that could not withstand the hurricane forces, and many penetrations through the levees for roads, railroads, and utilities. Furthermore, the levees were not designed to withstand overtopping.
- The hurricane protection system was designed for meteorological conditions (barometric pressure and wind speed, for example) that were not as severe as the Weather Bureau and National Weather Service listed as being characteristic of a major Gulf Coast hurricane.

As discussed in Chapter 2 for the Tohoku earthquake and tsunami, similar difficulties in hazard assessment and mitigation are common. In essence, they reflect failures to consider the problem as a whole and take a systems approach toward it.

One contributing factor to this failure to consider a problem as a whole is the tendency of various groups involved to consider only the aspects of the problem that concern them. As noted, various parts of the New Orleans protection system were built and operated by various agencies with different methods and goals. This problem also arises at national levels. In the US, for example, different agencies are concerned with different natural hazards – earthquakes, floods, storms, wildfire, drought, etc. Because each hazard is treated differently, there is no consistency. For example, the government's Federal Emergency Management Agency (FEMA) wants buildings built for the maximum wind expected once every 50 years (the typical life of a building); there is a 2% chance of this happening in any one year. However, they tell communities to plan for what is called the 100-year flood (section 8.1), which is a higher standard. This is the maximum flooding expected on average once every 100 years; i.e. there is a 1% chance of this happening in any one year. FEMA wants even higher standards for earthquakes: California should plan for the maximum earthquake shaking expected on average once in 500 years, and Midwestern states for the maximum shaking expected on average once in 2500 years. This pattern is the opposite of what one would logically do, because severe weather causes much more damage than earthquakes.

These examples illustrate another major problem, in that none of these policies come from careful analysis of their costs and benefits. It might better to plan for both 500-year floods and 500-year earthquakes. As we will see, using 2500 years is likely to overprepare for earthquakes. Conversely, it seems that in many areas planning only for the 100-year flood gives too low a level of protection, so it would be wise to prepare for larger floods. Similarly, the decision to rebuild the New Orleans protection system to withstand only a category 3 hurricane like Katrina, rather than larger storms, was made without considering the costs and benefits of the alternatives.

The lack of cost-benefit analysis sometimes reflects the fact that a government agency has no responsibility for the costs involved. Often, they create regulations whose costs fall on other governmental entities, such as local governments, or the private sector. Such unfunded mandates, based on the premise that someone else will pay, are often unrealistic.

Mitigation decisions are ultimately political, and so are influenced by bureaucratic inertia and special interests both inside and outside the governments involved. As the *New York Times* noted in discussing Japan's decision to rebuild tsunami defenses that failed in 2011,

> Some critics have long argued that the construction of seawalls was a mistaken, hubristic effort to control nature as well as the kind of wasteful public works project that successive Japanese governments used to reward politically connected companies in flush times and to try to kick-start a stagnant economy.

In the critics' view, the costly seawalls provide a false sense of security, and emphasis would be better placed on warning systems and improved evacuation procedures, which are far less expensive and can save many lives.

Analogous situations arise in formulating environmental policy, where the issue is often whether to permit or prevent activities, such as introduction of a new product, that may prove harmful. As for natural hazards, the question is what to do given that the hazard is poorly known, but may be large enough to justify action even given this uncertainty. Policies for such situation are often discussed in terms of the *precautionary principle*, a general concept implying that one should act to prevent possible harm even if it is unclear that an activity will be harmful. This concept has various formulations. In our view, the best is that of the United Nations 1992 Earth Summit, sometimes called the Rio Conference:

> In order to protect the environment, the precautionary approach shall be widely applied by States according to their capabilities. Where there are threats of serious or irreversible damage, lack of full scientific certainty shall not be used as a reason for postponing cost-effective measures to prevent environmental degradation.

This formulation recognizes that action may be needed although the hazard is uncertain, and that the costs and benefits of possible measures need to be considered in deciding how to act. Other formulations, however, neglect the cost issue, implying that one should make policies without considering their costs, which can lead to more harm than good. A classic example is that people should be vaccinated against disease only when the overall benefits outweigh the costs, including the risk of adverse reactions to the vaccine. As the 1976 swine flu fiasco (section 6.4) showed, sometimes the costs far outweigh the benefits.

The challenge is thus how to choose policies. In our view, the best way is to use an approach called *systems analysis* or *policy analysis*, which involves assessing a problem and exploring alternative solutions. It is used in defense planning, which is similar to natural hazard mitigation in that it involves defending society against threats that are hard to accurately assess. It is also used in business, which is similar to hazard mitigation in that both businesses and society seek to best use their resources for growth, and thus to mitigate losses due to external effects.

Systems analysis considers fundamental questions such as:

- What is the problem?
- What do we know and not know?
- What are we trying to accomplish?
- What strategies are available?
- What are the costs and benefits of each?
- What is an optimum – or at least sensible – strategy given various assumptions and uncertainty?

This process is not a set of formulas, but a way of thinking about problems. As Enthoven and Smith (1971) explain in their book *How Much is Enough?*, "the word 'systems' indicates that each decision should be considered in as broad a context as possible. The word 'analysis' emphasizes the need to reduce a complex problem to its component parts for analysis." It involves "asking the right questions, formulating the problem, gathering relevant data, determining their validity, and deciding on good assumptions." It seeks to replace "judgments made in a fog of inadequate and inaccurate data, unclear and undefined issues, conflicting personal opinions, and seat of the pants hunches" with ones made "in the clearer air of relevant analysis and experience, accurate information, and well-defined issues." Although it improves the decision-making process, it does not yield a single best solution, "because there is no universally valid set of assumptions. There are only better solutions and worse solutions."

7.2 Lessons from National Defense

During the Cold War, roughly spanning 1947–1991, the US and its allies feared a wide range of threats from the Soviet Union and its allies. As a result, the US defense budget grew enormously, averaging about 10% of the nation's gross domestic product (GDP), the total value of goods and services produced. (Present defense spending is about 5% of GDP.) The fact that these resources came at the expense of other needs was recognized. President Dwight Eisenhower stated, "Every gun that is made, every warship launched, every rocket fired signifies, in the final sense, a theft from those who hunger and are not fed, those who are cold and not clothed."

However, controlling the defense budget was a challenge. Instead of having a unified plan, each of the military services – army, navy, and air force – had its own goals and sought more funds for them. A joke of the time was that each service's military requirement was always 30% more than what it had. In the absence of a coherent plan, the army's Chief of Staff, General Maxwell Taylor, wrote: "It is not an exaggeration to say that nobody knows what we are really buying with any specific budget."

This situation changed when in 1961 President John Kennedy appointed Robert McNamara as Secretary of Defense. McNamara established a defense systems analysis group. His motivation – which applies equally to natural hazard mitigation – was set out as follows.

> The question of how to spend our defense dollars and how much to spend is a good deal more complicated than is often assumed. It cannot be assumed that a new weapon would really add to our national security, no matter how attractive the weapon can be made to seem, looked at by itself. Anyone who has been exposed to so-called 'brochuremanship' knows that even the most outlandish notions can be dressed up to look superficially attractive. You have to consider a very wide range of issues . . . You cannot make decisions simply by asking

yourself whether something might be nice to have. You have to make a judgment on how much is enough.

A crucial "how much is enough" question involved forces designed to attack the Soviet Union with nuclear weapons. The two sides were in an arms race, acquiring more and more very expensive long-range bombers and intercontinental missiles. A criterion proposed was that the US needed to have more destructive power, as measured in megatons (Mt) than the USSR. A megaton is a nuclear explosion releasing energy equal to a million tons of TNT, and is approximately the amount of energy released by a magnitude 7 earthquake. For comparison, the bomb that destroyed Hiroshima was much less than 1 MT, at 0.012 Mt.

However, it was recognized that neither side could win a nuclear war, because both would suffer enormous losses. Hence US strategy was to deter an attack by ensuring that the surviving US forces would inflict unacceptable damage on the Soviet Union after a Soviet attack. The systems analysis group estimated this damage, and found that beyond 400 Mt, little additional damage resulted (Table 7.1).

This result compares the cost of additional weapons to the "benefit" (a poor term in this context) of inflicting more damage. The benefit is the marginal return, or derivative, shown by the additional damage. As the number of megatons delivered increases, the marginal, or incremental, cost rises while the marginal damage flattens. Doubling the weapons from 400 to 800 Mt produces only a 1% "gain." Hence there was little to be gained by investing in additional weapons beyond 400 Mt, because the resources needed could be more usefully used otherwise.

Careful analysis also identified much less expensive ways of getting "enough." For example, rather than buying more fighter planes at $8 million each (in much lower 1960's prices) to make sure enough of them survived a

Table 7.1 Systems analysis: marginal return of nuclear warhead stockpile

Number of 1 Mt equivalent warheads	% Industrial capacity destroyed	% Marginal return
100	59	–
200	72	13
400	76	4
800	77	1
1200	77	0
1600	77	0

Adapted from Enthoven and Smith, 1971.

Soviet attack, the same result could be obtained by spending $100,000 per airplane for bombproof shelters.

Another "how much is enough" analysis is even closer to the natural hazard case, in that it involves probabilities. The US expected that countering a Soviet attack in Europe would require shipping troops and supplies across the Atlantic. Because the ships would have to be protected against Soviet submarines, the US and allies had to decide how much to invest in anti-submarine weapons.

The analysis considered the expected benefits of increasing the effectiveness of anti-submarine weapons, as measured by the probability of sinking a submarine and thus reducing the subsequent loss of ships. In probability theory, the expected value of an event is the value of the event times its probability (section 4.1). An analysis assumed that if a convoy of s ships were attacked by submarines that fire t torpedoes, each of which has probability r of sinking a ship, the expected value of the damage done by the submarines is rts. Similarly, if p is the probability that anti-submarine weapons will sink a submarine, the expected number of ships saved is $prts$.

The probability of sinking submarines could be increased by adding more weapon systems, for example minefields, US submarines, land-based patrol aircraft, carrier based aircraft, and destroyers. It could also be increased by making a given weapons system more effective, for example by improving sonar. However, these increases incur costs. How to invest can be analyzed by considering the derivatives of the benefit and cost curves, known as the marginal benefit and cost curves, which show the benefit and cost resulting from investing a little more. As shown in Figure 7.1, resources should be invested as long as the marginal benefit of increasing anti-submarine capability, $dp(rts)$, is greater than the marginal cost, $C'(v)$, where v measures the resources invested.

The marginal benefit curve is positive – greater than zero – because increasing anti-submarine capability saves ships. However, it flattens – has a negative second derivative – and gets closer and closer to zero with increasing v because once anti-submarine capability is well developed, improving it gets progressively harder because the most effective measures have already been taken. Similarly, the marginal cost rises for increasing anti-submarine capability, because at each level further improvements are harder to make.

The point where the margin benefit and margin cost curves cross is the optimum (point x in Figure 7.1). As long as the marginal cost is less than the marginal benefit, it makes sense to invest more. However, once the marginal cost is more than the marginal benefit, it no longer makes sense to invest

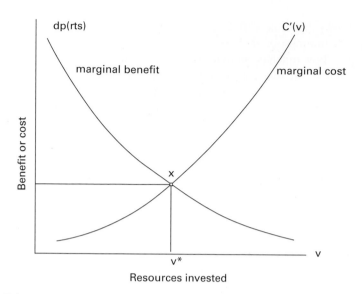

Figure 7.1   Comparison of the marginal benefit of increasing anti-submarine capability, *dp(rts)*, to the marginal cost *C'(v)*, as a function of the resources invested.

more. Although more ships would be saved, the resources required could do more good otherwise.

Graphs like this are schematic ways to guide our thinking, rather than functions we can compute exactly. Explicit calculations would involve assigning dollar values to the costs and benefits. Given the uncertainties involved in estimating the quantities involved, it would be unrealistic to claim to find an optimum strategy. However, even simple estimates can show which strategies make more sense than others. Thus although in real cases these approaches cannot give an optimum strategy, they can identify sensible strategies.

7.3   Making Choices

Systems analysis is also used to choose between options. A simple example would be a business that has two divisions, global technology (G) and software (S). The firm wants to allocate its total capital between the divisions to maximize its revenue *V*. If *a* is the fraction of capital allocated to sector G, then $(1 - a)$ is allocated to sector S, and the total revenue is

$$V = G(a) + S(1-a), \tag{7.1}$$

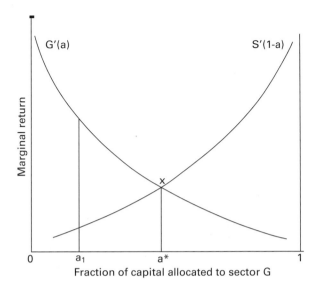

Figure 7.2 Marginal return on capital invested for sectors G and S. The optimum point x indicates that a* is the optimum allocation of capital between the sectors.

where $G(a)$ and $S(1 - a)$ are functions describing the revenue generated by the two sectors.

The optimal allocation between sectors can be found by considering the derivatives of the functions. Increasing the fraction a generates additional value by sector G of $G'(a)$. However, the resources taken away from sector S reduce the value it generates by $S'(1 - a)$. Thus $S'(1 - a)$ is the marginal cost of using an increment of capital in sector G.

This tradeoff is shown in Figure 7.2. If the current allocation is a_1, $G'(a_1)$ is greater than $S'(1 - a_1)$, and the value of the firm will increase by expanding sector G at the expense of sector S. At the optimum allocation point x, where the curves cross, $a = a^*$ and so:

$$G'(a^*) = S'(1 - a^*). \tag{7.2}$$

The gain to the value of the firm by increasing a from a_1 to a^* is the triangular area below the G′ curve and above the S′ one, given by the integral

$$\int_{a_1}^{a^*} [G'(a) - S'(a)]\, da.$$

7.4   Uncertainty and Risk Aversion

In natural hazard mitigation, the future is uncertain. Hence it is important to choose strategies that reflect uncertainty. An analogy for how to do this is a company trying to decide how much capital k to invest in a product that produces revenue $V(k)$. For an investment k the company expects an average, or mean, return V, but it recognizes that the actual return can range from $V(1 + h)$ to $V(1 - h)$.

For simplicity, consider only two outcomes, $V(1 + h)$ and $V(1 - h)$, such that $V(1 + h)$ is positive and $V(1 - h)$ is negative. In the first case the company makes money, and in the second it loses. The company estimates that there is a probability p, which is between 0 and 1, of outcome $V(1 + h)$ and a corresponding probability $q = 1 - p$ of outcome $V(1 - h)$.

Given this uncertainty, the company considers the expected value (EV) of its investment, the sum of the two outcomes weighted by their probabilities as:

$$EV = pV(1+h) + qV(1-h) = V + (p-q)h. \qquad (7.3)$$

If the two outcomes are equally likely, $p = q = 1/2$, so the expected value is just the mean value:

$$EV = V. \qquad (7.4)$$

However, the company is likely to be more concerned with not losing a given sum than in making the same sum, a situation called *risk aversion*. Risk aversion can be described mathematically as a case in which one will not accept a fair bet, one in which the chances of winning and losing are equal. For example, imagine you are invited to bet on flipping a coin, so you win $\$h$ if a head turns up, and lose the same amount if a tail turns up. Because the two outcomes are equally likely, this is a fair bet. If h was small, say \$1, you would probably accept the bet. However, if h was a large sum of money, you would be reluctant to take the bet, because you could lose a lot. You would only accept the bet if you win more, say $h + R$, if a head occurs than the h you would lose on a tail. The difference R needed for you to take the bet represents risk aversion.

The company can include risk aversion in deciding how much to invest, as shown in Figure 7.3. It wants to invest an amount of capital k that maximizes the net return function

$$Z(k) = V(k) - C(k), \qquad (7.5)$$

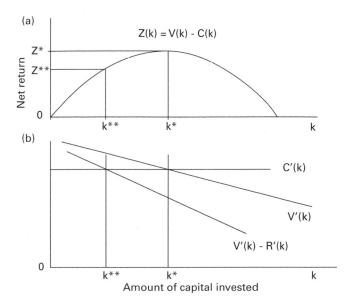

Figure 7.3 (a) In the absence of uncertainty and risk aversion, the maximum return less the cost of capital occurs for $Z(k^*) = V(k^*) - C(k^*)$. (b) This optimum occurs where the marginal return equals the marginal cost of capital, $V'(k^*) = C'(k^*)$. Including uncertainty and risk aversion reduces the optimum to k^{**}, given by $V'(k^{**}) - R'(k^{**}) = C'(k^{**})$.

where $V(k)$ is the return and $C(k)$ is the cost of this capital, which it has to borrow or divert from other applications. Due to diminishing returns to capital, $Z(k)$ is a concave function of k, with a maximum at $k = k^*$, where

$$V'(k^*) = C'(k^*), \qquad (7.6)$$

the marginal return equals the marginal cost. If there were no uncertainty or the company was not risk averse, it would invest this amount.

Because the company is risk averse, it also considers the range of possible outcomes for a given investment. The range about the mean value V is described by the variance σ^2, which reflects the probability of the different outcomes:

$$\sigma^2 = p[V(1+h) - V]^2 + q[V(1-h) - V]^2. \qquad (7.7)$$

If the two outcomes are equally likely, $p = q = 1/2$, the variance is

$$\sigma^2 = (hV)^2. \qquad (7.8)$$

Given this measure of the variability of the outcomes, the company describes how risk averse it is by multiplying the variance by a factor r, which gives a risk term

$$R = r\sigma^2. \tag{7.9}$$

R increases with the variance, showing that the larger the uncertainty in the outcome, the higher the risk. The company then maximizes a function that includes this term:

$$V(k) - C(k) - R(k). \tag{7.10}$$

As shown by the derivatives, this maximum occurs when

$$V'(k^{**}) - R'(k^{**}) = C'(k^{**}). \tag{7.11}$$

k^{**} is less than k^*, showing that the firm invests less because of the uncertainty in the outcome. The greater the risk term R, the less the firm invests.

7.5   Present and Future Value

In natural hazards planning, as in defense planning, society is using resources today to address possible future threats. It is thus important to consider the difference between the value of a sum today and its value in the future. If we take a sum of money today, $y(0)$, and invest such that it grows at an interest rate i, it will be worth

$$y(t) = y(0)(1+i)^t \tag{7.12}$$

at some future time t. This effect can be significant. An investment of $100 at 5% interest will be worth $432 in thirty years, and $13,151 in 100 years. This is because of *compounding* – we earn interest both on the original sum and on the interest already earned. As an old line says, the eighth wonder of the world is compound interest.

Conversely, the *present value* of a sum $y(t)$ at time t in the future is

$$y(0) = y(t)/(1+i)^t. \tag{7.13}$$

Thus assuming 5% interest, the present value of $100 thirty years from now is $23, and 100 years from now is only $0.77. In other words, $23 invested at 5% for thirty years would give $100, as would $0.77 for 100 years. The present value is how much we would invest today to get a certain sum at a future time.

These values are important because they have to be included when evaluating how much to spend today to reduce future damage. For example, it would not make sense to spend $50 today to avoid $100 damage a hundred years from now, because $50 is more than the present value of $100, which we have just seen is only $0.77. Investing the $50 would yield much more than $100. However, the issue is more complicated when considering reducing damage that can occur in any year between now and a future time.

The value assumed for the interest rate is crucial. This rate is also called the *discount rate*, because we discount the future by this rate. It reflects the tradeoff between the benefits of using resources today and those of reducing future damage. The lower the discount rate we assume, the more we need to invest today to earn a certain future sum. For example, assuming 1% interest, the present value of $100 thirty years from now is $74, compared to $23 for 5%. Thus the lower the discount rate, the more it makes sense to spend on mitigation. Conceptually, the question is whether society should invest in mitigation or some other investment – the less we can earn on the other investment, the more attractive mitigation becomes, and vice versa. We make similar choices in our personal lives. For example, given the choice between buying more car insurance or investing otherwise, we buy less car insurance if our investments are growing rapidly, because we will have more money to spend repairing or replacing the car if it is damaged. Because we do not know future interest rates, what we assume is crucial.

This issue has arisen in debates over the best policy to deal with the effects of the warming climate. Calculations using low discount rates favor more rapid reductions in carbon dioxide emissions compared to those using larger discount rates, which favor a longer ramp-up.

Typically, we deal with projects whose costs and benefits are spread out over time. For example, a hurricane protection system involves large initial costs and annual maintenance costs. The benefits, reduced losses if a major hurricane strikes, will occur in the future. Thus the costs and benefits can be compared using the present value of net benefits

$$PVNB = \sum_{t=0}^{T}(B_t - C_t)/(1+i)^t, \qquad (7.14)$$

where B_t and C_t are the benefits and costs in year t, i is the interest rate, and the summation runs from $t = 0$ to $t = T$.

For simplicity, assume that the major costs are incurred in year 0, and that from then on the annual costs and benefits are the same from year to year. Because we do not know when there will be benefits, we use the average rate of various major events, such as the rate of storms of different sizes, to find the probability that one will happen in any one year. This lets us find the expected benefit, the sum of the probabilities of an event of a given size times the anticipated loss reduction in such an event. Hence we can rewrite (7.14) by taking out the $t = 0$ term as

$$PVNB = -C_0 + (B_T - C_T)\sum_{t=1}^{T} 1/(1+i)^t = -C_0 + (B_T - C_T)D_T, \quad (7.15)$$

where C_0 is the initial cost, B_T and C_T are the average benefits and costs per year, and

$$D_T = 1/(1+i) + 1/(1+i)^2 + \ldots + 1/(1+i)^T = \sum_{t=1}^{T} 1/(1+i)^t \quad (7.16)$$

is the summation from $t = 1$ to $t = T$, representing the present value.

The summation is an example of a geometric series, which is a sum of a sequence of terms of increasing powers of a common ratio r:

$$S_n = 1 + r + r^2 + \ldots + r^{n-1} = \sum_{t=0}^{n-1} r^t. \quad (7.17)$$

The sum is found by a neat trick – multiply the series by r, giving

$$rS_n = r + r^2 + \ldots + r^n, \quad (7.18)$$

subtract the two series, which cancels everything but the first and last terms,

$$S_n - rS_n = S_n(1-r) = 1 - r^n \quad (7.19)$$

and divide both sides by $(1 - r)$, which shows that

$$S_n = (1 - r^n)/(1 - r). \quad (7.20)$$

In our case the common ratio is

$$r = 1/(1+i) \quad (7.21)$$

so the summation (7.17) becomes

$$S_n = (1 - (1/(1+i))^n) / (1 - 1/(1+i)) = ((1+i)^n - 1) / (i(1+i)^{n-1}). \quad (7.22)$$

The present value summation D_T differs from S_n in two ways. It has no $t = 0$ term, and goes up to the power $t = T = n$, whereas S_n goes only up to $n - 1$. Thus we can find D_T from S_n by adding $1/(1 + i)^n$ and subtracting one, which after a little algebra gives

$$D_T = ((1+i)^n - 1) / (i(1+i)^n) = ((1+i)^T - 1) / (i(1+i)^T) \quad (7.23)$$

where n now equals T, the total time period.

This summation can be large for long time spans. For $i = 0.05$, $D_T = 15.4$ for 30 years, and 19.8 for 100 years. For long enough times, the limit as T becomes infinite is

$$D_T = 1/i \quad (7.24)$$

so if $i = 0.5$, $D = 20$. Note that this is essentially the same as the value for 100 years.

The key point for hazard mitigation planning is that this summation reflects the decreasing present value of future benefits. The expected net benefit over 30 years, for 5% interest, is 15 times the annual benefit, not 30 times. Beyond 100 years, there is little and eventually no net increase in benefit, because the present value is so small. However, it still can make sense to prepare for events that might happen less often than once every 100 years, because they could happen next year or ten years from now.

7.6 Valuing Lives

A troubling but important point in mitigating hazards is that difficult choices involving lives often have to be cast in financial terms. The reason is that often the only way to compare the costs and benefits of various options – even in life and death applications – is to use a common financial metric. As Defense Secretary McNamara explained,

> I do not mean to suggest that we can measure national security in terms of dollars – you cannot price what is inherently priceless. But if we are to avoid talking in generalities, we must talk about dollars, for policy issues must sooner or later be expressed in the form of budget decisions on where to spend and how much.

Thus in the anti-submarine warfare example, Enthoven and Smith criticize the argument for spending more because it would save some sailors' lives if war came:

> This, unfortunately, is the kind of emotional argument that often frustrates defense planning. Surely if the proponents were really concerned about human life, they would be looking for ways of spending that $2 billion on things such as improved health or medical facilities in the US and overseas, or food for starving people, or even cheaper ways of saving lives through naval forces. It is hard to imagine an easier task than to find ways of spending $2 billion which would be more productive in saving lives.

One way to approach these issues is to compare the cost of saving lives in various ways. Although these estimates have uncertainties, often the differences are large enough to show that some methods are more cost-effective than others. For example, it has been estimated that seat belts in cars cost about $30,000 for each life saved, while air bags cost about $1.8 million for each life saved.

These relative comparisons avoid explicitly putting values on lives. Still, in many applications there is no way to avoid valuing lives. We value our own lives when we decide how much life insurance to buy. Hence as adults get older, they typically reduce the amount of life insurance they carry. We also value our lives by deciding how much to spend on safety features in cars or other products we buy. For example, if we will pay $50 but no more for a feature that has a 1/100,000 chance of saving our lives, we are implicitly valuing our lives at $5 million. Similarly, for better or worse, hospitals, insurance companies, and government agencies decide that some medical treatments are too expensive. In the same way, mitigating hazards ultimately comes down to how much society is willing to spend to save lives.

Society values lives in two ways. One is through the legal system, where lives are valued by the present value of what people would earn. Thus after the September 11, 2001 attacks, when the US government compensated the families of those killed, payments averaged $2 million, but were based on the victim's income. This meant that families of the poorest paid received $250,000, whereas those of the highest paid received more that $7 million. Although this approach is common, it values people differently and awards the most to the wealthiest, raising ethical questions.

Another approach is to look at how much society has been willing to pay in the past to save lives. The numbers vary, because our willingness to spend money reflects preferences beyond what is purely logical. Much more is spent trying to protect against scary exotic risks, such as rare diseases or terrorism, than on much greater but familiar risks such as traffic accidents or common diseases. Only about a third of US adults get shots to protect against the flu that kills about 35,000 people every year. On the other hand, it is not clear that any lives are saved by the US Transportation Security Administration, which has 60,000 employees and a budget of $8 billion, mostly to screen airline passengers who are very unlikely to be terrorists. In fact, researchers find that the inconvenience and unpleasantness of airport security costs lives, because some people chose to drive – which is far more dangerous – rather than fly.

Typically, the value used by government agencies to decide whether to require a life-saving measure is about $5 million per life. Thus a proposal to require mattresses less likely to catch fire was estimated to cost $343 million, but was viewed favorably because it was estimated to save 270 lives per year. In contrast, seat belts are not required on school busses because they were estimated to save one life per year at a cost of $40 million.

7.7 Implications for Natural Hazard Mitigation

Although for natural hazards we rarely know enough to do detailed calculations, this chapter illustrates approaches that can be applied. These involve assumptions and thus do not yield unique or precise answers. However, they provide an organized way of thinking about a problem that considers various options and is likely to yield better decisions.

Put another way, graphs like that in Figure 7.1 are schematic ways to guide our thinking, rather than functions we can compute exactly. It is conceptually useful to think about optimum strategies. However, given the simplifying assumptions made and the uncertainties involved in estimating the quantities involved, it would be unrealistic to think that we can actually find an optimum strategy. Still, even simple estimates can show which strategies make more sense than others. Thus although in real cases these approaches cannot give an *optimum* strategy, they can identify *sensible* strategies.

Some aspects of the approach are worth noting.

One is *the need to think about the entire system*, rather than isolated aspects. For example, considering whether building levees to protect communities from river flooding makes economic sense requires also considering the extent to which the levees promote further growth in vulnerable areas, which will eventually flood. Moreover, the levees raise water levels, so floods downstream will be higher. Sensible policy requires considering these factors together.

Another is that *costs and benefits are hard to estimate precisely*. It is hard enough to estimate the value of the financial quantities involved, but even harder to value lives or intangibles such as the esthetic cost of building a huge seawall along an attractive shore. Different people and societies value things differently. Moreover, some people's costs are other people's benefits. Money spent on a seawall is a cost to society as a whole, but a benefit to the contractors who build it.

A third point is *the concept of marginal cost and benefits*. So long as the marginal benefit, the expected reduction in losses due to more mitigation, is greater than the marginal cost of the additional benefit, more mitigation makes sense. However, there is a level of mitigation beyond which we gain little benefit even if we spend a great deal more on mitigation. It makes sense to increase mitigation levels until this point, but increasing them further diverts resources that could do more good otherwise.

A fourth is *the importance of estimating the expected benefit of mitigation*, which is the expected reduction in loss. Overestimating or underestimating the expected loss can happen either because we inaccurately assess the hazard, which affects the probability of loss, or how effective mitigation measures will be, which affects the magnitude of the loss. As we have seen, the resulting uncertainties can cause major problems.

Finally, despite its potential value in formulating policy, systems analysis is only useful if *those making policy are willing to ask key questions and consider alternatives*. If those in charge are committed already to a policy – which is often the case in government – demonstrating its weaknesses is likely to have no effect.

For example, Secretary of Defense McNamara accepted his analysis group's recommendations that the cost of a number of proposed weapons systems was much greater than their benefit, and so decided not to build them. However, he was a "hawk" committed to US involvement in the Vietnam war, and thus had no interest in systems analysis of issues relating to it. Presumably, he knew that the results would not be what he wanted to hear. In particular, US strategy was based on wearing down the enemy. However, the systems analysis group's study found that most enemy losses occurred when it chose to fight, so that the enemy could control its casualty rate and thus keep it at a sustainable level. Hence, the analysis concluded correctly, "the notion that we can win the war by driving the enemy from the country or inflicting an unacceptable rate of casualties is false."

Similarly, in dealing with natural hazards, society can make rational policy, if it chooses to.

Questions

7.1. Motor vehicle accidents cause over 30,000 deaths per year in the US This number has been steadily decreasing for more than forty years.

How would you decide if this is happening as a result of accident mitigation or just because people drive less?

7.2. Possible strategies to reduce the number of deaths in motor vehicle accidents even further include lowering speed limits, improving roads, making cars safer, and reducing drunk and otherwise distracted driving, including drivers using cellular telephones and sending text messages. For each of these approaches, consider its costs and benefits, some of which are economic and others are noneconomic, and illustrate them with simple graphs. How would you decide what combination to adopt?

7.3. In the US, bicycle riders are encouraged and often required to wear helmets. In contrast, in Europe – where bicycling is more common and safer – few adults wear helmets. European health experts argue that requiring helmets leads people to view a basically safe activity as dangerous, and thus discourages it. In their view, the small increased risk of not wearing helmets is greatly outweighed by the resulting reduction in obesity and heart disease. This issue is increasingly being considered because cities promoting bicycle-sharing programs are finding that they are much more successful if helmets are not required. How would you analyze this issue?

7.4. The US Transportation Security Administration has about 60,000 employees and a budget of $8 billion. Some argue that this cost and the resulting delays and inconvenience to the traveling public are justified. Others regard the program as pointless "security theater" and argue that the resources used could be better used otherwise. Moreover, airport security costs lives in that some people are deterred from flying and instead drive. How would you analyze this issue?

7.5. A long-standing debate at universities is how to allocate the budget between science departments that bring in research grants and humanities departments that teach more students. If you were the president of your university, what rational way would you use to allocate the educational budget among the various departments? Develop a method and explain how you would justify it to the entire faculty.

7.6 Following the September 11, 2001 attacks in New York and Washington, the US government compensated the families of those killed. Payments averaged $2 million, but were based on the victim's income, so families of the poorest paid received $250,000, whereas those of the highest paid received more that $7 million. The program raised many questions. Those favoring it argued that it prevented lawsuits against the airlines, whereas opponents pointed out that in general the government does not compensate families of victims of crimes, natural disasters, or accidents. Moreover, was it sensible or fair to pay the most to families of the wealthiest victims, who were far more affluent than the average taxpayer? Suggest and justify a policy for dealing with possible similar events in the future.

7.7.   Imagine that the community you live in has received a one-time increase in revenue from the federal government equal to 1% of its annual budget. Develop a method to allocate this money between education, police and fire services, recreation, public works, disaster mitigation, and any other needs you think appropriate. Explain how you would justify this allocation to the community's government.

7.8.   For a 3% interest rate, calculate how much $100 invested today will be worth in thirty years, and the present value of $100 thirty years from now.

7.9.   Derive equation 7.23 from equation 7.22 and equation 7.24 from equation 7.23.

7.10. Hazard mitigation policymaking involves deciding how much resources to invest for possible future gain. A somewhat analogous issue is parents investing in their childrens' sports in hopes of winning a college athletic scholarship. To explore this issue, imagine that you are the parent of a sixth grader. To estimate the costs involved, take the average annual cost parents report spending per child in grades 6–12, $670, and use equation (7.23) to compute its present value. To estimate the benefit, assume that the child will play sports in high school, which gives him or her a 2% chance of an athletic scholarship whose value averages $10,000 per year, and use equation (7.13) to find its present value. From a financial point of view, how wise an investment is this? What conditions would make the result different?

Further Reading and Sources

The "hubristic" quote on tsunami defense rebuilding is from Onishi (2011). UNESCO (2005) reviews the precautionary principle. The Taylor and McNamara quotations are from Enthoven and Smith (1971).

Baumol (1961) gives an overview of policy analysis methods that discusses the optimization techniques used by Enthoven and Smith. Manski (2013) illustrates applications to current policy issues, and Fuchs (2011) discusses health care applications. Morgan and Henrion (1990) discuss uncertainty issues in policy analysis. Nordhaus (2008) discusses the role of the discount rate in climate policy.

The air bag and seat belt cost effectiveness data are from Levitt and Dubner (2011). September 11 compensation data come from *http://money.cnn.com/2011/09/06/news/economy/911_compensation_fund*. Singer (2009) explains how government agencies value lives. Rosenthal (2012) explores the bicycle helmet issue.

Athletic scholarships and children's sport costs are discussed by Pennington (2008), Heitner (2012), and O'Shaughnessy (2012).

References

American Society of Civil Engineers Hurricane Katrina External Review Panel, *The New Orleans Hurricane Protection System: What Went Wrong and Why*, American Society of Civil Engineers, Reston VA, 2006.

Baumol, W., *Economic Theory and Operations Analysis*, Prentice-Hall, Englewood Cliffs, NJ, 1961.

Enthoven, A., and K. Smith, *How Much is Enough? Shaping the Defense Program 1961–1969*, Rand Corporation, Santa Monica, CA, 1971.

Fuchs, V., *Who Shall Live?* World Scientific, Singapore, 2011.

Heitner, D., 1 In 5 American parents spending more than $1,000 per child on sports related expenses, http://www.forbes.com/sites/darrenheitner/2012/10/04/1-in-5-american-parents-spending-more-than-1000-per-child-on-sports-related-expenses/, 2012.

Levitt, S., and S. Dubner, *SuperFreakonomics*, William Morrow, New York, 2011.

Manski, C. F., *Public Policy in an Uncertain World*, Harvard University Press, Cambridge, MA, 2013.

Morgan, G. M., and M. Henrion, *Uncertainty: A Guide to Dealing with Uncertainty in Quantitative Risk and Policy Analysis*, Cambridge Univ. Press, Cambridge, 1990.

Nordhaus, W., *A Question of Balance: Weighing the Options on Global Warming Policies*, Yale. Univ. Press, New Haven, CT, 2008.

Onishi, N., Seawalls offered little protection against tsunami's crushing waves, *New York Times*, March 13, 2011.

O'Shaughnessy, L., 8 things you should know about sports scholarships, http://www.cbsnews.com/8301-505145_162-57516273/8-things-you-should-know-about-sports-scholarships/, 2012.

Pennington, B., Expectations lose to reality of sports scholarships, *New York Times*, March 10, 2008.

Rosenthal, E., To encourage biking, cities lose the helmets, *New York Times*, September 29, 2012.

Singer, P., Why we must ration health care, *New York Times magazine*, July 19, 38–43, 2009.

UNESCO, United Nations Educational, Scientific and Cultural Organization, *The Precautionary Principle*, 2005.

8

Guessing the Odds

"All models are wrong. Some models are useful."

George Box, statistics pioneer[1]

8.1 Big Events Are Rare

Researchers interested in various natural hazards have spent a lot of time trying to understand the probabilities of events. Although the specific approaches vary for different hazards, the key result is that the biggest and potentially most destructive events are the rarest. As a result, their probabilities are the hardest to estimate from the limited historical records available. Doing this involves a lot of uncertainty, which we need to keep in mind. Let's explore these issues, building on the basic ideas about probability introduced in Chapter 4.

Figure 8.1 shows an example for floods of the Red River of the North in Fargo, North Dakota. The data are the maximum flow in the river for each year from 1882–1994. The maximum flow in most years was about 5,000 cubic feet per second (cfs). However about once every ten years on average the flow reached about 12,000 cfs, and occasionally even larger floods occurred. Hydrologists used the data to derive the flood frequency plot shown, comparing the maximum annual flow to the recurrence time, the average time between floods of that size. Using a logarithmic scale for the recurrence time, the data are well fitted by a straight line.

[1]George Box, statistics pioneer. Box and Draper, 1987. Reproduced with permission of John Wiley & Sons.

Playing against Nature: Integrating Science and Economics to Mitigate Natural Hazards in an Uncertain World, First Edition. Seth Stein and Jerome Stein.
© 2014 John Wiley & Sons, Ltd. Published 2014 by John Wiley & Sons, Ltd.
Companion Website: www.wiley.com/go/stein/nature

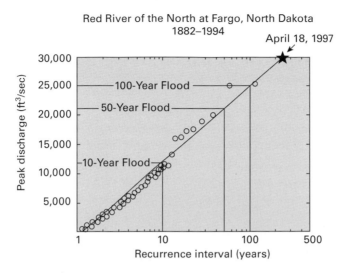

Figure 8.1 Estimation of flood frequency from a long-term record. (Eric Baer, SERC.)

The flow levels that occur on average every 10, 50, and 100 years are called the 10-year, 50-year, and 100-year floods. Put another way, these are the flow levels that in any one year we expect that there is a 10% (1/10), 2% (1/50), or 1% (1/100) chance of occurring. However, the floods do not come regularly, so there can be a 10-year flood in two successive years.

Physically, that is because the weather is variable, so the maximum rainfall in one year does not depend much on what it was last year. There are wet years and dry years, so there are years with high flow and years with low flow. Large deviations are rarer, so larger floods are rarer.

Mathematically, this is described by saying that the 100-year flood has a probability of 1%, or $p = 0.01$ of happening in any one year. The probability that it will not happen is $q = 1 - p$ because $p + q = 1$, since we are certain ($p = 1$) that either it will happen or it will not.

To figure out how likely that flood is to occur within a time period, we assume that what happens in one year does not depend on what happened in previous years. In the language of probability, this means that each year's largest flood is an *independent event*. As we discussed in section 4.2, the probability of multiple independent events all happening is their product.

Thus the probability that this flood will happen in year $n + 1$, after n years in which there wasn't one, is pq^n. This lets us find the total probability of having a flood sometime in n years: it is the sum of the probability that it happened in the first year, plus the probability that it did not happen in the first year but happened in the second, plus the probability that it did not in the first two years but happened in the third, etc.

$$P(n, p) = p + pq + pq^2 + pq^3 + \ldots + pq^{n-1} = p(1 + q + q^2 + q^3 + \ldots + q^{n-1}).$$

$$(8.1)$$

The notation $P(n,p)$ denotes the total, or cumulative, probability that something whose probability of happening in each year is p will happen at least once in n years.

Equation 8.1 is a geometric series with common ratio q, so using equation (7.20) gives its sum as

$$P(n, p) = p(1 - q^n)/(1 - q) = 1 - q^n.$$

$$(8.2)$$

This makes sense – as we showed in section 4.2, the probability that in n years there will be at least one such flood is one minus the probability that there will not be one. This formulation – the probability of something happening is one minus the probability that it will not – is very useful in considering many kinds of natural disasters.

Let's look at specific examples. The 100-year flood, by definition, has a 1% chance of happening in any one year, and a 99% chance of not happening, so $p = 0.01$, $q = 0.99$.

For a period of thirty years, the typical length of a home mortgage, $n = 30$, and equation (8.2) gives $P = 0.26$, so there is a 26% chance of experiencing the 100-year flood.

Interestingly, for $n = 100$, the equation gives $P = 0.63$, so in 100 years there is only a 63% chance of experiencing the 100-year flood. Conversely, there is a 37% chance that it will not happen.

This may seem strange, because our instinct might be to say if the probability is 1%, multiply $n = 100$ years by $p = 0.01$, which gives $np = 1$, so it is sure to happen. This does not work because although these floods are 100 years apart on average, they are not regular. Think of tossing a coin – on average heads comes up once every two tosses, but in two tosses the probability of at least one head is one minus the probability of no heads: $1 - (1/2)(1/2) = 3/4$, not 1. As we noted in section 4.2, often we think something is "sure to happen," but it really is not.

To show why just using np often does not work, we rewrite equation (8.2) in terms of the probability p,

$$P(n, p) = 1 - q^n = 1 - (1 - p)^n,$$

$$(8.3)$$

and expand $(1 - p)^n$ using a Taylor series. The Taylor series formula says that the value of a function $f(x)$ near a point a is

$$f(x) = f(a) + f'(a)(x - a) + f''(a)(x - a)^2/2 + f'''(a)(x - a)^3/6 + \ldots \quad (8.4)$$

where $f'(a)$ is the first derivative, $f''(a)$ is the second derivative, and so on. In this case, $f(x)$ is x^n, $x = 1 - p$, $a = 1$, and $(x - a) = -p$, so $f(1) = 1$, $f'(1) = n$, $f''(n) = n(n - 1)$, etc. Hence

$$(1 - p)^n = 1 - np + n(n - 1)p^2/2 - n(n - 1)(n - 2)p^3/6 + \ldots \quad (8.5)$$

and

$$P(n, p) = 1 - (1 - p)^n = np - n(n - 1)p^2/2 + n(n - 1)(n - 2)p^3/6 + \ldots \quad (8.6)$$

The expansion shows that the probability is np plus an alternating series of terms. We can only neglect these other terms when np is small compared to one. For example, approximating the probability of a 100-year flood in 10 years using $np = 10\ (0.01) = 0.10$ matches the exact $1 - (0.99)^{10} = 0.10$. However, for np large, we need the other terms. For example,

$$P(100, 0.01) = 1 - 0.495 + 0.167 - 0.092 = 0.628, \quad (8.7)$$

which is quite different from $np = 1$. We need the next three terms to get the correct answer.

Several problems can arise. First, the longer a time history is available, the better we can estimate the recurrence times. However, the large events are rare, so in statistical terms we're taking a small sample from a population described by an unknown distribution. The issue is that although the sample mean is an unbiased estimate of the population mean, equation 4.31 showed that the standard deviation of the sample mean is equal to the standard deviation of the population divided by the square root of the sample size. Because the sample of large events is small, the standard deviation – the uncertainty – of the recurrence time estimate is large.

Second, estimating how often really big events will happen often involves extrapolating beyond the limits of the observations. Extrapolating suggests that the big flood of 1997 should occur on average once in 250 years. However, extrapolating outside the available observations can cause problems.

Third, this analysis assumes that the probability of floods has not changed with time. On a timescale of a few years that is often approximately true, but not always. The probability can change in several ways. Figure 8.2 shows two cases where human activity changed flood patterns. For Mercer Creek, river flows are higher because the area became more built up, so the ground absorbs less rain and more water runs into the creek. For the Green River, building a dam upstream reduced river flow. In the first case the estimated 100-year flood level increased, and in the second it decreased.

Rainfall and thus floods also change as the climate changes. For example, long- term rainfall records can be derived from tree rings showing how much a tree grew each year, because the growth depends on rainfall. Records going back more than 1200 years in the southwestern US show 100-year long periods of high and low rainfall. Climate models predict changes in rain patterns as the climate warms, so human activity will likely make the probabilities of floods time-dependent. This seems to be occurring in Europe, where winter storm tracks from the Atlantic are shifting northward, causing increased rain and flooding in northern Europe and increasing drought in southern Europe. In such cases, the available record of rain and flooding will likely not correctly describe the future, so insight is being sought from climate models. A challenge is that global models have even greater uncertainty when they are "downscaled" to describe how regional climate will change.

8.2   Time-Independent Probability Models

In the last section, we considered the probability of a flood of a certain size, which is a discrete event that either happens or does not. We can take this analysis further using an analogy. A traditional model for the probabilities of events is a container or urn containing a number of balls (Figure 8.3): e balls are labeled "E" for event, and n balls are labeled "N" for no event. In each time period, drawing a ball shows whether an event will happen. The probability of an event, written $P(E)$, is the probability of drawing an E-ball. This is the ratio of the number of E-balls to the total number of balls,

$$P(E) = p = e/(e+n), \tag{8.8}$$

and the probability of no event is the ratio of the number of N-balls to the total number of balls:

$$P(N) = q = n/(e+n). \tag{8.9}$$

Both p and q are less than one, as probabilities always are, and their sum

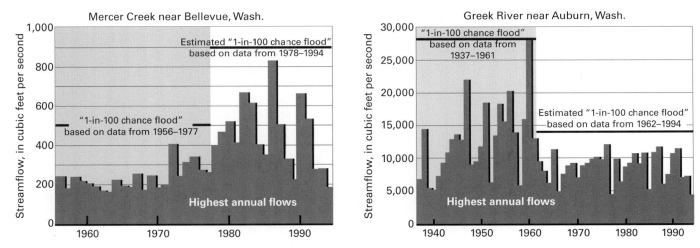

Figure 8.2 Changes in flood frequency due to human activity. (Dinicola, 1996.)

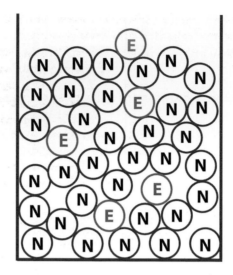

Figure 8.3   A model for the probability of an event is drawing a ball from an urn filled with balls, some labeled "E" for event and others labeled "N" for none. (Stein and Stein, 2013a. Reproduced with permission of the American Geophysical Union.)

$$p + q = e/(e + n) + n/(e + n) = (e + n)/(e + n) = 1 \qquad (8.10)$$

is one because an event is certain to either happen or not. In mathematics an event is often called "success" and a non-event "failure", but for hurricanes, volcanoes, floods, or earthquakes these may not be good terms, because people want "failure".

The simplest case is called sampling with replacement – after we draw a ball from the urn, we put it back in. As a result, the probability of an event in each successive draw is constant or *time-independent*. Successive draws are independent events because the outcome of one does not change the probability of what will happen in the next. Put another way, the system has no "memory."

We express this mathematically by saying that the joint probability $P(AB)$ that two independent events A and B both occur is the product of the probabilities of each occurring (equation 4.16). We used this idea in the last section, when we assumed that the largest flood in an area in a given year is independent of what happened in previous years. Similarly, because hurricane tracks in a given year often cross each other (Figure 3.9), it looks as though the fact that an area was struck by one storm does not prevent another from striking it in the same year. As a result, hurricane hazard models use time-independent probability.

   This lets us see how the time between events – or waiting time – evolves. We use the fact that the probability of the event happening in k draws is one minus the probability that it will not, as we derived previously in equation (8.2):

$$P(k, p) = 1 - q^k. \tag{8.11}$$

Because q is less than one, then as k gets large, $P(k,p)$ approaches one. Thus even though the probability of an event is p in every draw, the more draws we have, the more likely it becomes that an event has happened. The higher p is, the lower q is, and the shorter a time we have to wait for the probability of an event to reach a given value.

   To estimate the probability of more than one event, we use the *binomial probability distribution*:

$$B(m, n, p) = C_{n,m} p^m q^{n-m} = \frac{n!}{m!(n-m)!} p^m q^{n-m}, \tag{8.12}$$

which is the probability that in n trials (or draws) there will be m events and $n - m$ non-events. As before, p and q are the probabilities of an event and a non-event. The term $C_{n,m}$ is the number of ways we can have m events and $n - m$ non-events. It is written in terms of factorials, where $n! = n \times (n - 1) \times (n - 2) \ldots \times 3 \times 2 \times 1$ and $0! = 1$. For example, in three trials we can get one event and two non-events in $C_{3,1} = 3!/(1!2!) = 3$ ways: ENN, NEN, or NNE, so we multiply $p^1 q^2$ by 3.

   The binomial distribution is complicated to compute, so an approximation is used when the number of trials n is large and the probability p of an event is small. In this case, because $n \gg m$,

$$\frac{n!}{(n-m)!} = \frac{n(n-1)\ldots(n-m+1)(n-m)!}{(n-m)!} = n(n-1)\ldots(n-m+1) \approx n^m.$$

$$\tag{8.13}$$

Because p is small, using the Taylor series in equation 8.5 gives

$$q^{n-m} = (1-p)^{n-m} = 1 - p(n-m) + p^2(n-m)(n-m-1)/2! + \ldots$$
$$\approx 1 - pn + (pn)^2/2! + \ldots \approx e^{-np}. \tag{8.14}$$

These approximations let us replace the binomial distribution by another probability distribution that is easier to compute, called a *Poisson distribution*:

$$P(m, n, p) = n^m p^m e^{-np}/m! = (np)^m e^{-np}/m!. \tag{8.15}$$

The Poisson distribution can be written in terms of the probability of having m events in a time period t if on average they occur τ years apart. Because the expected number of events in the n trials is $np = t/\tau$,

$$P(m, t, \tau) = (t/\tau)^m e^{-t/\tau}/m! \tag{8.16}$$

The Poisson distribution is really a function of only two variables, the number of events m and the mean number expected $\mu = np = t/\tau$, so it is sometimes written as

$$P(m, \mu) = \mu^m e^{-\mu}/m!. \tag{8.17}$$

Poisson distributions are often used in hazard assessments. For example, between the years 1722 and 2005, 45 hurricanes are thought to have struck the central Gulf of Mexico coast, including New Orleans. These occurred on average 6.3 years apart, so the average number per year was 0.16. A Poisson model predicts that the probability that no hurricanes will strike in a year is $P(0, 0.16) = e^{-0.16} = 0.85$ or 85%, the probability of one is $P(1, 0.16) = 0.16\, e^{-0.16} = 0.136$ or 14%, and the probability of two is $P(2, 0.16) = (0.16)^2\, e^{-0.16}/2 = 0.011$ or 1%. The probability at least one will strike is one minus the probability that none will, 1– 0.85 or 15%.

In general, the probability of at least one event in t years if on average they occur τ years apart is 1 minus the probability of no events, so

$$P(m > 0, t, \tau) = 1 - P(0, t, \tau) = 1 - e^{-t/\tau} \approx t/\tau, \tag{8.18}$$

where the last step used the first term of the Taylor series expansion $e^{-x} \approx 1 - x$ and so is valid for $t \ll \tau$, or a time interval much shorter than the average time between events.

Thus if a large earthquake strikes an area on average every 100 years, a Poisson model predicts that the probability of one occurring in the next 10 years is $1 - e^{-10/100} = 0.095$, which is close to 10/100 or 0.1. However, estimating the probability of one occurring in the next 50 years requires using the full term $1 - e^{-50/100} = 0.39$, which differs from 50/100 = 0.5 because 50 years is a large fraction of the average time, 100 years.

This Poisson analysis gives a very similar result to that for the waiting time between events (8.11). Although the term $1 - e^{-t/\tau}$ looks different from $(1 - q^k)$, both give very similar results. As we saw, if an event happens on average every 100 years, in 50 years the waiting time analysis gives the chance of at least one event as $(1 - 0.99^{50}) = 0.39$ and the Poisson analysis gives $1 - e^{-50/100} = 0.39$.

As before, this equation shows that in a time period equal to the average time between events, the probability of having at least one event is $1 - e^{-1} = 0.63$ or 63%. That is large, but not one. The fact that something happens on average once every τ years does not mean that in that time it is guaranteed to happen.

It is important to remember that these probabilities are time-independent – the system has no memory. Whatever has or has not happened in the past, the probability of an event in the next trial does not change. It does not matter when the last event happened, so *events cannot be overdue*. The good news is that no matter how long it is been since a coastal town was last hit by a hurricane, it is no more likely to be hit next year. The bad news is that having been hit last year does not make it less likely this year. It is like tossing coins – the fact that the last four tosses came up tails does not make heads more likely.

Assuming that an event that has not happened recently is overdue is called the "gambler's fallacy," because people have lost lots of money that way. A famous case is that between 2003 and 2005, the number 53 had not come up in Venice's lottery. Assuming that it was overdue, Italians bet more that 3.5 billion Euros on it, and lost. Some ran up debts, some went bankrupt, and four people died in 53-related incidents before 53 eventually showed up.

8.3   Time-Dependent Probability Models

In some applications, we expect that the probabilities of events will change with time. For example, the idea of an earthquake cycle is that strain accumulates across a fault and is released in earthquakes (Figure 2.2), so we might expect the probability of a large earthquake to decrease after one occurs, and then rise.

We can use the urn model to describe a time-dependent process by supposing that after drawing a ball we do not replace it. If the first draw is an N-ball, the probability of an E-ball in the second draw is now

$$P(E) = e/(e+n-1), \tag{8.19}$$

because there is one less N-ball in the urn. This number is bigger than the probability p of drawing an E-ball in the first draw, because we are dividing by a smaller number. Drawing an N-ball on the first draw thus makes drawing an E-ball in the second more likely. In this case, the probability is variable or *time-dependent*.

We can model a general time-dependent process by adding a number of E-balls to the urn after a draw when an event does not occur, and removing a number of E-balls when an event occurs. This makes the probability of an event increase with time until one happens, after which it decreases and then grows again. Events are not independent, because one happening changes the probability of another.

To see how this works, imagine that an event has just happened, leaving e E-balls and n N-balls in the urn. On the next draw, if no event happens, we add a number a of E-balls to the urn. That increases the probability of an event in the next trial from its initial value $p = e/(e + n)$ to a new value

$$P(E|N) = (e + a)/(e + a + n), \tag{8.20}$$

which is the conditional probability (section 4.2) of E in this draw given N in the last draw. From then on, after each draw in which an event does not happen, an additional a E-balls are added. After k non-events the conditional probability of drawing an E-ball is

$$P(E|kN) = (e + ka)/(e + ka + n). \tag{8.21}$$

With time, as k rises, the probability rises from its initial value p to approach one if no E-ball has yet been drawn.

Eventually, an event occurs. We then remove r E-balls, which decreases the probability of an event in the next draw to

$$P(E|E) = (e + ka - r)/(e + ka - r + n), \tag{8.22}$$

where $P(E|E)$ is the conditional probability of E in this draw given E in the last draw.

This approach is well suited for computer modeling, because we can use random number generators to simulate sampling from a distribution. Figure 8.4 shows an example for an urn that initially has 20 E-balls and 380 N-balls, giving 1/20 initial probability of an event, or on average one every 20 years for samples that are drawn yearly. If the probability is time-independent ($a = r = 0$), this probablility of 1/20 will not change, as shown by the straight line.

The situation differs for time-dependent models, as shown by two cases. In one case (shown by the long-dashed line), with $a = 1$ and $r = 20$, the probability growth between events is approximately offset by the decrease after events, so on average the probability oscillates about the time-independent

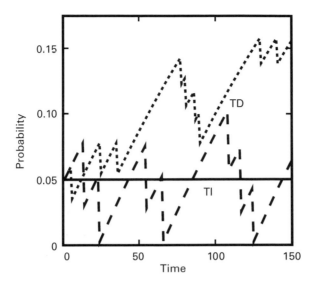

Figure 8.4   Comparison of the probability of an event as a function of time for a time-independent (solid line) and two time-dependent (dashed lines) urn models. (Stein and Stein, 2013a. Reproduced with permission of the American Geophysical Union.)

case. In the other case (shown by the short-dashed line), with $a = 1$ and $r = 10$, the probability decreases less after events and so tends to increase. The sequences of events that arise result from both the model parameters and chance, so another run with the same parameters, or one with different parameters, could yield a very different-looking sequence.

This kind of formulation can be used to model specific time-dependent processes by choosing criteria for how balls are added and removed. The values of a and r control the average time between events and the variation of the actual times around this average, so we can have models that are both strongly and weakly time-dependent. For example, we might describe the probabilities of hurricanes as mostly time-independent with a time-dependent component due to climate change. We could also make a version of the model that allowed for events of different sizes.

For natural hazards, we do not know the model or the parameters. Instead, we try to infer these from the history of past events and ideas about the process. This is like trying to infer the contents of the urn and the sampling process from the samples that have already been drawn. Naturally, going from a sequence of events to what model generated them is very hard. The challenge is illustrated by the sequence of events (Figure 8.5) in the model runs shown in Figure 8.4. The upper sequence comes from the time-independent

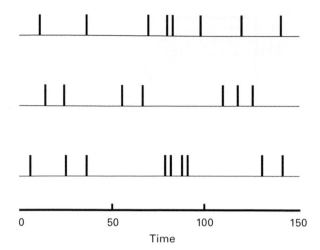

Figure 8.5   Sequence of events as a function of time for the time-independent (top line) and time-dependent (lower lines) urn model runs in Figure 8.4. (Stein and Stein, 2013a. Reproduced with permission of the American Geophysical Union.)

model, and the other two sequences result from the time-dependent models. The differences among the three reflect both the different models and the effects of randomness.

Looking at these sequences, it would be difficult to tell what process generated them. Moreover, we can easily convince ourselves that we see all sorts of patterns, many of which would look different in another run, as illustrated by the differences between the lower two series that are due only to randomness. Some parts of the sequences look pretty regular, so if we had only these samples, we might decide the system was periodic, and then would be surprised when the next event did not fit that apparent pattern. We might then decide that something in the system gave clusters of periodic events separated by longer intervals, and would be disappointed when that pattern also broke down. People are very good at identifying apparent patterns in nature – some of which prove to be real, and some of which do not.

Researchers have used various models to describe the recurrence of natural hazards. In addition to Poisson models that yield time-independent behavior, expected time-dependent behavior is modeled by assuming that the time between events is described by a Gaussian or other probability density function. The resulting uncertainties in predicting when events will occur are shallow if the model and parameters approximate reality reasonably well, and deep if they do not. In the first case we can give the probability of a future event with reasonable confidence, whereas in the second we cannot. Our best way of distinguishing between the two is to examine how well models predict what occurs. We still face the problem that although a model has done a reasonable job of matching the sequence of past events, it may not do well for future ones if it does not adequately describe the underlying process.

Questions

8.1. Using a spreadsheet or computer program, use equation 8.2 to calculate how many years it takes for there to be a 95% probability of experiencing a 100-year flood.

8.2. A coastal area has not been struck by a hurricane in 28 out of 46 years. Assuming that the incidence of hurricanes is described by a Poisson distribution (equation 8.17), estimate its mean parameter μ. In how many years would you expect one hurricane?

8.3. Give an example you have encountered of the "gambler's fallacy" and explain why it is wrong.

8.4. Of the two time-dependent probability models in Figure 8.4, which looks more like an earthquake cycle model and why? What physical situation might the other describe?

8.5. In March 2012, Britain's Meterological Office told the government:

> The forecast for average UK rainfall slightly favours drier than average conditions for April-May-June, and slightly favours April being the driest of the three months.

Water companies prepared for water shortages. Later, the Meterological Office admitted that,

> given that April was the wettest since detailed records began in 1910 and the April-May-June quarter was also the wettest, this advice was not helpful

and its chief scientist stated

> The probabilistic forecast can be considered as somewhat like a form guide for a horse race. It provides an insight into which outcomes are most likely, although in some cases there is a broad spread of outcomes, analogous to a race in which there is no strong favourite. Just as any of the horses in the race could win the race, any of the outcomes could occur, but some are more likely than others.

How do you respond to these statements? What – if anything – would you suggest doing differently?

8.6.   The synthetic time histories of earthquakes in the two lower panels of Figure 8.5 result from the same physical model, and differ only due to randomness. Imagine, however, that these were actual time histories of magnitude 6 earthquakes in different areas, and the 150-year time is today. Based on these data, how would you describe the earthquake history to residents of each area? What would you tell them about the earthquake hazard they face today?

8.7.   Synthetic time histories of earthquakes or other events, like those in Figure 8.5, can be generated using a computer program or spreadsheet with a random number generator. For example, Excel's RAND() function returns a random number evenly distributed between 0 and 1. Write a program or spreadsheet for a sequence of 200 steps in which at each step there is a 1/20 chance of an event occurring. Consider what patterns appear, what caused them, and what you might infer if you had the sequence but no information about how it was generated.

Further Reading and Sources

Figure 8.1 is from *http://serc.carleton.edu/quantskills/methods/quantlit/floods.html*. Kirby (1969) and Klotzbach and Gray (2010) discuss flood and hurricane probabilities. The history of the "53" lottery number is described by Arie (2005). The Meteorological Office's "not helpful" forecast is described by BBC News on March 29, 2013 *http://www.bbc.co.uk/news/science-environment-21967190*.

References

Arie, S., No 53 puts Italy out of its lottery agony, *The Guardian*, February 10, 2005.

Dinicola, K., The 100-Year Flood, *US Geological Survey Fact Sheet 229-96*, 1996.

Kirby, W., On the random occurrence of major floods, *Water Resour. Res.*, 5, 1641–1648, doi: 10.1029/WR005i004p00778, 1969.

Klotzbach, P., and W. Gray, United States hurricane landfall probability, report, Colorado State University, 2010.

Stein, S., and J. L. Stein, Shallow versus deep uncertainties in natural hazard assessments, *Eos Trans. AGU*, 94, 133–134, 2013a.

9

When's the Next Earthquake?

"With four parameters I can fit an elephant, and with five I can make him wiggle his trunk."

John von Neumann, mathematician

9.1 A Very Tough Problem

Unlike the situation for hurricanes, it is not clear whether the recurrence of large earthquakes is better described as time-independent or time-dependent. Sorting this out from the limited data is hard, because big earthquakes on a given fault are hundreds or thousands of years apart. We rarely have more than one observed seismologically, so we have to depend on historical accounts and geological records of paleoearthquakes, known as paleoseismic records. As a result, earthquake histories have a variety of biases. Moreover, even with a good paleoseismic history, the times between earthquakes often have a more complicated pattern than we would expect from simple probability models.

We are thus trying to make sense of a process whose physics we do not fully understand with a small historical sample. This makes it often hard to tell which apparent patterns are real and which are artifacts of the small sample. The failed Parkfield earthquake prediction (section 3.3) showed that we can fit a model to a time history, but it may or may not be useful in describing what happens next.

Let's see what we can and cannot do.

Playing against Nature: Integrating Science and Economics to Mitigate Natural Hazards in an Uncertain World, First Edition. Seth Stein and Jerome Stein.
© 2014 John Wiley & Sons, Ltd. Published 2014 by John Wiley & Sons, Ltd.
Companion Website: www.wiley.com/go/stein/nature

9.2   Earthquake Frequency-Magnitude Relation

The simplest approach is to assume time-independence, and to try to estimate earthquake recurrence from the statistics of earthquakes.

The starting point is that the number of earthquakes that occur in a given area depends on magnitude, with successively smaller earthquakes being more common. This variation is described approximately by the *earthquake frequency-magnitude*, or *Gutenberg–Richter* relation

$$\log N = a - bM, \tag{9.1}$$

where N is the number of earthquakes with magnitude greater than M that occurred in a given time. This relation is similar to that in Figure 8.1, showing that larger floods are rarer.

Although the intercept, a, depends on the number of earthquakes in the time and region sampled, the slope, b, is generally about 1, as shown in Figure 9.1 for earthquakes around the world. Thus there is an approximately tenfold

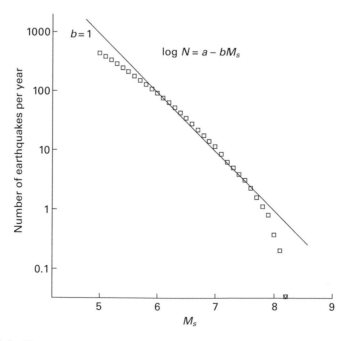

Figure 9.1   Frequency-magnitude plot for ~13,000 earthquakes with surface wave magnitude $M_s \geq 5.0$ during 1968–1997. The line shown, with slope b about 1, fits the data reasonably well. (Stein and Wysession, 2003. Reproduced with permission of John Wiley & Sons.)

increase in the number of earthquakes for successively smaller magnitudes. Equivalently, the average time between earthquakes of a given magnitude is about ten times longer than that for earthquakes one magnitude smaller.

The Gutenberg–Richter relation also applies in individual seismic areas, with *b* generally about 1. Thus although the number of earthquakes depends on how seismically active an area is, the relative frequency still applies. For example, in the period from about AD 700 to 2003, Japan is estimated to have had about 190 earthquakes with M > 7 and 20 with M > 9. Similarly, in the period 1816–2003, southern California had about 180 earthquakes with M > 6 and 25 with M > 7, whereas the New Madrid seismic zone in the central US had about 16 earthquakes with M > 5 and 2 with M > 6. Although the precise numbers, especially for the rarer large earthquakes, depend on the period chosen and uncertainties in estimating magnitudes prior to the invention of the seismometer in about 1890, the logarithmic decay still appears.

Frequency-magnitude plots let us extrapolate observations of small earthquakes to say something about the rare big ones. However, there are a number of problems. Often our observations of the rate of big earthquakes do not match what the rate of small ones implies. This could mean several things. It could indicate that we just have not seen enough big ones, due to the short earthquake history available. It could indicate a problem in comparing the magnitudes and rates of big earthquakes inferred from paleoseismology with those measured seismologically. It could also indicate that the area in question does not have earthquakes above a certain size.

For example, since Japan has had 20 magnitude 8 earthquakes in 1300 years, there's on average one every 65 years. Hence if magnitude 9 earthquakes occur, there should be one on average somewhere in Japan about every 650 years. This analysis does not say whether there actually are magnitude 9 earthquakes, something which was unclear before March 2011. Now that we know they happen, we also suspect they could happen on any portion of the subduction zone that is long enough for a long rupture (section 2.2), so it is hard to say anything useful about how often they should happen on any particular part, such as the Nankai Trough.

To see how the frequency-magnitude relation relates to slip on faults, we need to think physically about magnitudes. Earthquake magnitude, introduced in 1935 by Charles Richter, is based on the fact that the amplitude of ground motion recorded by a seismometer reflects the earthquake's size once it is corrected for the decrease in amplitude with distance from the earthquake. Different magnitudes are labeled by "M" with a subscript showing what kind they are. A common one, the surface wave magnitude M_s, is based on the size of seismic waves that travel near the earth's surface. This is calculated using the formula

1906 San Francisco Earthquake (M$_w$ = 7.9) seismogram

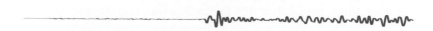

2004 Sumatra Earthquake (M$_w$ = 9.3) seismogram

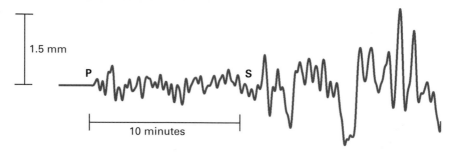

1.5 mm

P

S

10 minutes

Figure 9.2 Comparison of seismograms for the 1906 San Francisco and 2004 Sumatra earthquakes, shown on the same scale. (Richard Aster, Colorado State University. Reproduced with permission.)

$$M_s = \log(A/T) + 1.66 \log d + 3.3, \tag{9.2}$$

where A is the amplitude of the wave, T is the period of that wave – usually 20 seconds – and d is the distance from the seismometer. "Log" is the base 10 logarithm, so a one unit increase in magnitude, as from 5 to 6, indicates a tenfold increase in seismic wave amplitude.

Figure 9.2 shows the ground motion from the 1906 San Francisco earthquake recorded in Europe and that recorded at about the same distance from the giant 2004 Sumatra earthquake. The motion from the 2004 earthquake is about 2.5 mm, corresponding to magnitude 9.3. Motion from the 1906 earthquake, which had magnitude 7.8, is about ten times smaller.

Earthquakes range from tiny, which have negative magnitudes, to giant, with magnitude 9. A magnitude 9 earthquake causes ground motion ten million times larger than that of a magnitude 1 earthquake. Using the logarithmic scale makes earthquake sizes easier to understand. There are words to describe magnitudes. Magnitude 3 earthquakes are minor, 4 are light, 5 are moderate, 6 are strong, 7 are major, and 8 or higher are great. The biggest since the invention of the seismometer was the magnitude 9.5 earthquake in 1960 along the Chilean coast.

As shown in Figure 9.3, for the whole world in a typical year there is about 1 magnitude 8 earthquake, about 17 of magnitude 7, and so on. The exact

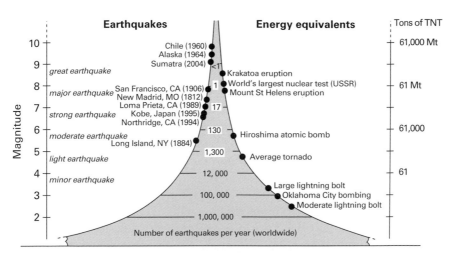

Figure 9.3 Comparison of earthquakes with different magnitudes in terms of how often they happen and the energy they release. (Stein and Wysession, 2003. Reproduced with permission of John Wiley & Sons.)

numbers vary from year to year but the approximate factor of ten relation is clear. Most of the energy is released in the few largest earthquakes. Often, the biggest earthquake in a given year releases more energy than all of the others that year put together. Thus although there are millions of earthquakes every year, only a few large ones do much harm.

These numbers give a reality check when people claim to be predicting earthquakes. Usually they do no better than just guessing based on the average. For example, a magnitude 7 earthquake occurs somewhere in the world about every month, and a magnitude 6 occurs about every three days. In 1990, a much-touted Midwest earthquake prediction was said to be credible because its author claimed to have predicted the 1989 Loma Prieta earthquake. In fact, he had actually said that near that date there would be an earthquake somewhere in the world with magnitude 6, which was a safe but useless bet.

A natural question is why the Gutenberg–Richter relation works – why are big earthquakes less common? The answer is not obvious, because magnitudes were defined before anyone knew much about earthquakes, and so do not directly reflect what actually happened during an earthquake.

Intuitively, we would expect that bigger earthquakes occur because larger areas of a fault slipped for longer distances. Earthquake physics shows this to be right. The energy measured in the longest period waves recorded on seismograms is proportional to a property of an earthquake called the *seismic moment*, written M_o:

$$M_o = \mu SD = \text{fault rigidity} \times \text{fault area} \times \text{slip distance}. \qquad (9.3)$$

We have a good idea of μ, the fault's rigidity or strength, both from earthquake studies and from lab experiments on rocks. The area of the fault that slipped, S, which we approximate as a planar surface, can be estimated from seismograms or from the locations of aftershocks, which map out the area of the fault that moved. How far the fault slipped, D, can be estimated from seismograms or by dividing the measured seismic moment by the rigidity and fault area. If the earthquake ruptured the earth's surface, these values can be checked by measuring the length of surface rupture and the amount of offset across it.

However, because seismic moments have units of energy like dyne-cm and are large numbers, they are hard to explain to the media and public. Thus seismologists use the moment magnitude M_w that is calculated from the seismic moment using the relation

$$M_w = (\log M_o / 1.5) - 10.73. \qquad (9.4)$$

The constants in this equation were picked so that the moment magnitude approximately matches other magnitudes, so the moment magnitude is now called "the" magnitude.

The moment magnitude lets us relate the amount of ground motion to what happened on the fault. To see how this works, look at the comparisons in Figure 9.4. The boxes show the approximate area of the fault that broke in each earthquake, which is the length along the fault times the width, which is the distance measured down the fault plane. They also show the amount of slip on the fault, which is an average because some parts slip more than others.

The smallest earthquake shown is the 1994 magnitude 6.7 Northridge, California, earthquake. In that earthquake, an area about 15 km on a side slipped an average distance of about a meter. Although the earthquake was not that big, it occurred below the heavily populated Los Angeles area, and so caused 58 deaths and $20 billion of damage, which is a large loss for an earthquake of this size. The next earthquake shown is the magnitude 6.9 Loma Prieta earthquake, in which an area of the San Andreas fault about 40 km long and 15 km wide slipped about 2 meters. These two are a lot smaller than the 1906 San Francisco earthquake, in which a part of the San Andreas about 450 km long and 10 km wide slipped about 4 m.

Although in California people often talk about the next big San Francisco earthquake – the recurrence of the 1906 one – as "the big one", the 2011 Tohoku earthquake was almost unbelievably bigger. In it, an area about 400 km long and 200 km wide slipped about 20 m on average, with some parts slipping much further.

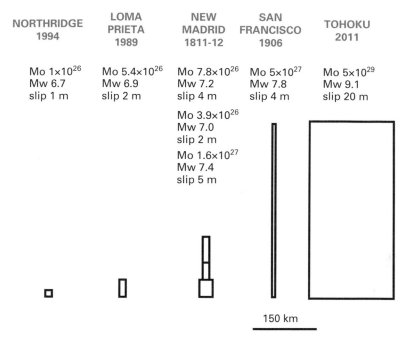

Figure 9.4 Comparison of rupture areas and slip distances for earthquakes with different magnitudes. Seismic moments are given in dyne-cm.

The seismic moment and magnitude of earthquakes derived from seismograms, and hence the values inferred for fault area and slip distance, have significant uncertainties. The final value reported is an average of results from seismometers in different places. These results differ because of differences in how well seismic waves travel through different parts of the earth. In addition, different methods give somewhat different results. Studies of the Loma Prieta earthquake found magnitudes that vary by 0.2 units. Older data cause even greater uncertainties. The magnitude of the 1906 San Francisco earthquake was once thought to be 9.3 but is now thought to be about 7.8, and the fault length has been estimated as anywhere between 300 to 500 km. Things are even iffier when discussing earthquakes such as those of 1811–12 in New Madrid, which occurred before the seismometer was invented. The values shown come in part from historical descriptions of the shaking and damage, and so have large uncertainties.

The seismic moment lets us see why the Gutenberg–Richter relation arises. As we discussed in section 3.4 when talking about chaos, we can think of big earthquakes as growing at random from small earthquakes. In that case, the probability of an earthquake of a given size on a fault is inversely proportional to the area of the fault involved in the earthquake. The fault area is reflected in the seismic moment and thus the magnitude, so larger magnitude earthquakes are less likely and thus rarer.

9.3   Earthquake Cycle Model

The Gutenberg–Richter curve only gives us an idea about how often on average an earthquake of a given magnitude will happen on a fault. That is useful, but it would be good to do better. For example, knowing that a magnitude 8 earthquake happens on a given fault on average every 150 years is somewhat useful, but we could make much better mitigation policies with more specific information.

As a result, lots of effort has gone into trying to estimate time-dependent probabilities to describe earthquake recurrence. The idea comes from the earthquake cycle model, in which strain on faults accumulates slowly over time and is released eventually in large earthquakes. Let's motivate it geologically and then look at how well it works.

Most large earthquakes reflect the motions between the plates of the lithosphere, the cold strong rocks that form the earth's outer shell, which is about 100 km thick. The boundaries between the plates come in three types, which are simplest when they are in the rocks under the oceans (Figure 9.5). At divergent

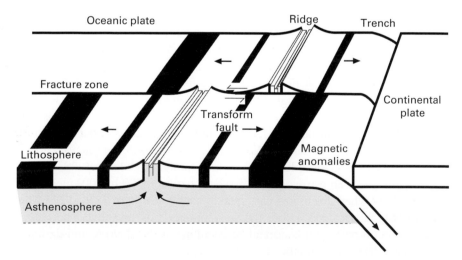

Figure 9.5   Types of plate boundaries in oceanic lithosphere. Oceanic lithosphere is formed at ridges and subducted at trenches. At transform faults, plates slide by each other. (Stein and Wysession, 2003. Reproduced with permission of John Wiley & Sons.)

boundaries plates move away from each other, whereas at convergent boundaries they move toward each other. At the third boundary type, transform faults, plates slide by each other. Divergent oceanic boundaries, known as mid-ocean ridges or spreading centers, are long lines of underwater volcanoes. Here hot rock upwells, cools, and is added to the two plates. The material continues cooling as the plates move away from the ridges. It eventually reaches convergent boundaries known as subduction zones or trenches. Here the plate is consumed as it descends back into the mantle, reheating as it goes, and producing volcanoes on the upper plate. The three boundary types also exist in the continents, but are more complicated.

Because most large earthquakes result from motions between plates, a map of earthquake locations nicely shows the plate boundaries (Figure 9.6). As we have seen, the 2011 Tohoku earthquake occurred at the Japan Trench, the subduction boundary between the subducting Pacific plate and the overriding plate, which is either part of the North American plate or a smaller Okhotsk plate (Figure 2.1). The 2010 Haiti earthquake (Figure 1.4) resulted from motion on a transform fault between the North American and Caribbean plates. Some plate boundaries, typically those under the oceans, are narrow, whereas others in the continents are wide zones. For example, the boundary zone between the Pacific and North America plates in the western US is more than 1000 km wide, from the west coast as far east as Utah. Although most of the motion between the two plates occurs across the San Andreas fault, the rest is distributed across the zone, causing earthquakes in places such as Nevada.

There are also some, but far fewer, large earthquakes within plates. The three magnitude ~7 earthquakes that occurred near New Madrid, Missouri, in 1811 and 1812 are examples of such *intraplate* earthquakes.

We know the rates and directions of plate motions pretty well. We have observations of rates and directions of spreading at midocean ridges over the past few million years, recorded by seafloor magnetic anomalies and the directions of transform faults. Combining these observations with the direction of motion in earthquakes at plate boundaries gives models of plate motions worldwide (Figure 9.7). We can also measure plate motions over the past few years directly using GPS and other space geodetic methods.

To see how these motions give rise to earthquakes, let's look at the San Andreas fault, part of the boundary between the Pacific and North American plates. Early in the morning of April 18, 1906, geologist G. K. Gilbert had a long-awaited experience. Although he had studied faults in the field, he had never felt a major earthquake, and so explained,

> It is the natural and legitimate ambition of a properly constituted geologist to see a glacier, witness an eruption, and feel an earthquake . . . When, therefore

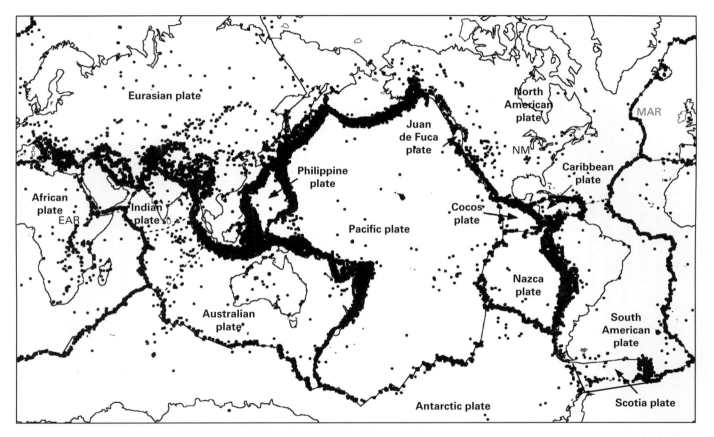

Figure 9.6 Map of major plates and earthquake locations, shown by dots. The earthquakes outline most plate boundaries. "NM" marks New Madrid. "MAR" is Mid-Atlantic Ridge. "EAR" marks the East African Rift. (Stein and Wysession, 2003. Reproduced with permission of John Wiley & Sons.)

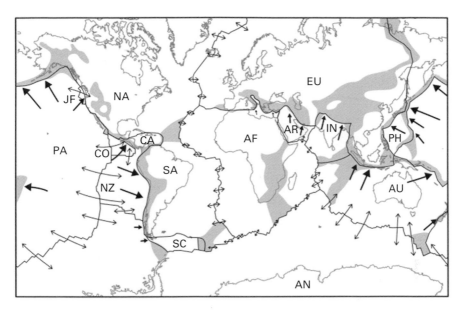

Figure 9.7 Major plates and the relative motion at their boundaries. Arrow lengths show the speed of the motion. Diverging arrows show spreading at mid-ocean ridges. Single arrows on a subducting plate show convergence. Stippled areas are diffuse plate boundary zones. (Gordon and Stein, 1992. Reproduced with permission of American Association for the Advancement of Science.)

> I was awakened in Berkeley on the eighteenth of April last by a tumult of motions and noises, it was with unalloyed pleasure that I became aware that a vigorous earthquake was in progress.

The earthquake and resulting fires did enormous damage, especially in San Francisco. About 3,000 people are thought to have died and about $10 billion (in today's dollars) of damage occurred. Gilbert and colleagues, who formed an investigating commission, studied the earthquake's effects and found a fascinating pattern. At many places, the ground was broken (Figure 9.8). For about 450 km along the zone now known as the San Andreas fault, features like fences, roads, or train tracks that used to be straight had been offset, with their west sides moved north relative to their east side by about four meters.

Commission member Harry Reid used these observations to propose the *elastic rebound* or *earthquake cycle* theory used to explain earthquakes on a fault. Between earthquakes, rocks on opposite sides of the fault move in opposite directions. However, friction on the fault "locks" it, so the fault cannot move, while strain accumulates in a zone on either side of the fault

Figure 9.8 Some results of the earthquake on the San Andreas fault, April 18, 1906. (a) Ground breakage along the fault trace. (b) A fence offset by the earthquake. (Gilbert, 1906. Reproduced with permission of the U.S. Geological Survey.)

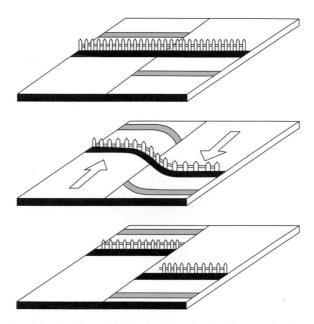

Figure 9.9 How elastic rebound works is shown by the history of a fence across a fault. (Stein and Wysession, 2003. Reproduced with permission of John Wiley & Sons.)

that deforms elastically. Eventually the strain overcomes the friction and the fault slips in a large earthquake, by a distance long enough to "catch up" with the motion away from the fault that accumulated over the years since the previous large earthquake (Figure 9.9). Thus the earthquake releases the accumulated elastic strain.

Another phrase for this behavior is "stick-slip": the fault is stuck for a long time, and then slips. This can be illustrated by attaching a rubber band to a soap bar in a box, and pulling it across a yoga mat (Figure 9.10). Because of friction between the box and the mat, the box does not move while the rubber band stretches. As the rubber band stretches, it applies increasing force to the box. Eventually, the force overcomes the friction, the rubber band snaps back to its original length, and the box suddenly slips forward.

Why the rocks on opposite sides move was not understood until the discovery of plate tectonics in the 1960s explained that the San Andreas fault is the boundary between the Pacific and North American plates. Most faults on which large earthquakes occur were quickly identified as plate boundaries.

GPS data let us observe the earthquake cycle between earthquakes (Figure 9.11). A GPS receiver uses radio signals sent from satellites orbiting above the earth. Using the fact that radio waves travel at the speed of light, the receiver

Figure 9.10   An analogy for elastic rebound: using a rubber band to pull a soap bar in a box across a mat.

figures how far away each satellite is, and where the receiver has to be for each satellite to be that distance away. Because GPS measures the positions of markers on the ground very precisely, measurements over time give the speed of ground motion.

Figure 9.11 shows the rate of motion of GPS sites across the Carrizo plain segment of the San Andreas. The farthest western sites move northward at a rate of 36 mm/yr relative to the farthest eastern site. Sites in the strained zone between them show a smooth variation across the locked fault – in theory an arctangent function – like that shown by the fence in the center panel of Figure 9.9. The elastic strain that is accumulating will be released in a future large earthquake.

The GPS data, which show the rate of motion over a few years, can be compared to geological data that record the motion over long times. A stream called Wallace Creek crosses the San Andreas south of the part of the fault that broke in 1906 (Figure 9.12). In the past, the creek ran straight across. But as the Pacific plate moved to the left (north) relative to the North American plate, it shifted the creek so the parts on either side of the fault are now about 130 m apart. Radiometric dating of charcoal in the streambed shows that this happened over the past 3700 years. Thus over this time, the average speed of plate motion here is 130 m / 3700 years, or 35 millimeters per year, so the rate of motion averaged over the past few thousand years is the same as we see today.

The rate of motion across the San Andreas and the amount of slip in large earthquakes gives a way to estimate how often these earthquakes happen. Although the plates move at this speed, motion on the fault itself only happens

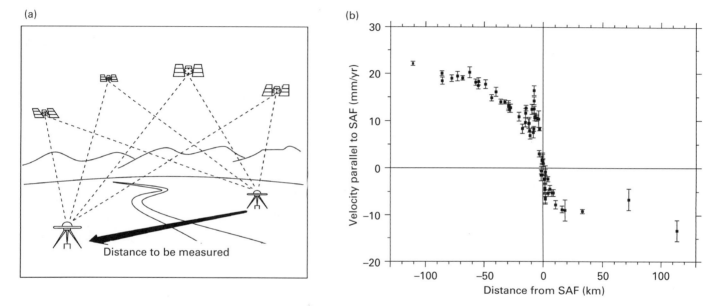

Figure 9.11 (a) Very precise plate velocities can be obtained by measuring the positions of receivers over time using GPS signals. (b) Profile across the San Andreas showing GPS velocities in the direction along the fault. (Stein and Wysession, 2003. Reproduced with permission of John Wiley & Sons.)

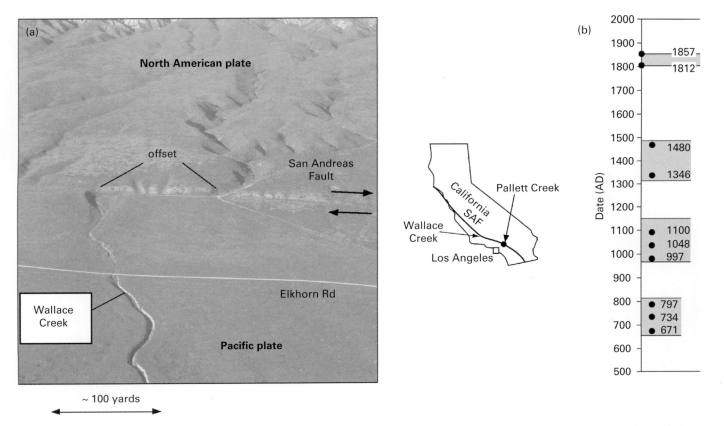

Figure 9.12 Geological studies of the San Andreas fault. (a) The rate of motion across the fault, and thus between the Pacific and North American plates, is measured using the offset of Wallace Creek. (b) Paleoearthquake history at Pallett Creek. (Stein, 2010. Reproduced with permission of Columbia University Press.)

in big earthquakes. In 1906 the fault slipped about 4 m. Accumulating this at 36 mm/yr would take 4000 mm / 36 mm/yr = 111 years. So if the big earthquakes on this part of the San Andreas are like that of 1906, they should be about 110 years apart.

This estimate is consistent with the record of paleoearthquakes at a nearby site on the San Andreas called Pallett Creek, using sand layers that were disturbed by earthquake shaking. The most recent earthquake was in 1857, when historical records show that an earthquake about the size of the 1906 one happened. As shown in Figure 9.12, earlier large earthquakes appear to have occurred approximately 132 years apart. Because 1857 plus 132 years is 1989, maybe an earthquake is overdue.

However, this earthquake history also shows that the interval between earthquakes is variable, varying from 45 years to 332 years with a standard deviation of 105 years. Earthquakes seem to come in clusters: three between 671 and 797, then a 200 year gap, then three between 997 and 1100, followed by a 246 year gap. Thus although we might have expected the next one 132 years after 1857, or in 1989, it is not surprising that it did not happen. If the recent cluster is still going on, it may happen soon. On the other hand, if the cluster that included the 1812 and 1857 earthquakes is over, it may be a long time until the next big earthquake.

There are other complexities. The dates have uncertainties, some earthquakes may have been missed, and some earthquakes may have been bigger than others. Thus although such data give an average time between earthquakes, they cannot predict the next one very well. For example, the authors of a study in 1989 estimated that the probability of a major earthquake before 2019 was somewhere between 7% and 51%, and the real uncertainty may be more.

The fact that these recurrence times are so variable is striking because the study used a long history of ten earthquake cycles, which is longer than we have in most places. Imagine having only part of that history – a history within one cluster would make the recurrence look more regular than it is, and one spanning two clusters could make the recurrence look even less regular.

Highly variable recurrence is what would be expected from the chaos theory idea (section 3.4): the plate motion controls the average rate of large earthquakes, but the intervals between them are variable. This complicated earthquake history looks a lot like the histories produced by the urn models in Figures 8.4 and 8.5. That example showed that simple models with random elements can generate complicated histories, and that it is very hard to use those histories to figure out what is actually going on. This is a major problem for hazard assessment.

9.4   Computing Earthquake Probabilities

What we have discussed gives us two ways to estimate the probability of future earthquakes.

The simplest is to use the time-independent model from section 8.2, in which earthquake probabilities satisfy a Poisson distribution. As we saw in equation 8.18, the probability of one or more events in t years is 1 minus the probability of no events,

$$P(m > 0, t, \tau) = 1 - P(0, t, \tau) = 1 - e^{-t/\tau} \approx t/\tau, \qquad (9.5)$$

where the last step is valid for $t \ll \tau$, or a time interval much shorter than the average time between events. Thus if a large earthquake strikes an area on average every 100 years, the probability of one in the next 10 years is $1 - e^{-10/100} = 0.095$, which is close to 10/100 or 0.1. In this model, the probability that an earthquake will occur in an interval of time starting from now does not depend on when "now" is, because a Poisson process has no "memory." On average, earthquakes are separated by time τ, but when the last one occurred has no effect.

The Poisson model is not appealing, because our seismological instincts favor earthquake cycle models, in which strain builds up slowly after a major earthquake to produce the next one. In this case, the probability of a large earthquake should be small immediately after a large earthquake, and then grow with time. This is described by time-dependent models in which the probability of a large earthquake a time t after the past one is given by a probability density function $p(t, \tau, \sigma)$ that depends on the average and variability of the recurrence times, described by the mean τ and the standard deviation σ.

In other words, p gives the probability that the recurrence time for this earthquake will be t, given an assumed distribution of recurrence times. The cumulative probability that the earthquake will occur by time T since the past earthquake is found by integrating the density function

$$P(T) = \int_0^T p(t, \tau, \sigma)dt. \qquad (9.6)$$

We want to estimate how likely the occurrence of an earthquake is between now and some future time. This is the conditional probability that the earthquake will occur between time T_0 (now) and a future time T, given the condition that it has not yet happened by time T_0. We find this using Bayes' theorem (equation 4.14),

$$P(A|B) = P(AB)/P(B), \tag{9.7}$$

in which $P(A|B)$ is the conditional probability $C(T, T_0)$ that the earthquake will occur between T_0 and T, $P(AB)$ is the joint probability that it will occur in that interval, and $P(B)$ is the probability that it hasn't yet happened by T_0, which is just 1 minus the probability that it has. Hence:

$$C(T, T_0) = (P(T) - P(T_0)) / (1 - P(T_0)). \tag{9.8}$$

The denominator is less than one, so the conditional probability is greater than the joint probability (the numerator), showing that the fact that the earthquake hasn't happened yet makes it more likely. This increase shows that the probability is time-dependent.

This approach can be used with any assumed probability density function. The simplest is to assume that earthquake recurrence times are described by the familiar Gaussian or normal distribution (section 4.3) whose probability density function is

$$p(t, \tau, \sigma) = \frac{1}{\sigma(2\pi)^{1/2}} \exp\left[\frac{-1}{2}\left(\frac{t - \tau}{\sigma}\right)^2\right]. \tag{9.9}$$

This distribution is often described using the normalized variable $z = (t - \tau)/\sigma$, which describes how far t is from its mean, in terms of the standard deviation.

Figure 9.13 shows such an analysis for the segment of the San Andreas fault including the Pallett Creek site (Figure 9.12), on which the last major earthquake was in 1857. The analysis uses a Gaussian distribution with a mean and standard deviation of 132 and 105 years, whose probability density function is shown. These are used to estimate the conditional probability that a major earthquake would occur between 2013 and 2033. These times are 156 and 176 years since 1857, and so correspond to normalized times of 0.23 and 0.42, with probabilities of 0.59 and 0.66. Thus the conditional probability is

$$\begin{aligned} C(2033, 2013) &= (P(2033) - P(2013)) / (1 - P(2013)) \\ &= (0.66 - 0.59) / (1.0 - 0.59) = 0.18, \end{aligned} \tag{9.10}$$

or 18%. The probability for successive twenty-year intervals increases with time, and so is 20% if the earthquake has not occurred by 2033, and 22% if it has not occurred by 2053.

It is interesting to compare these time-dependent probabilities to those predicted by the time-independent Poisson model. For an assumed mean

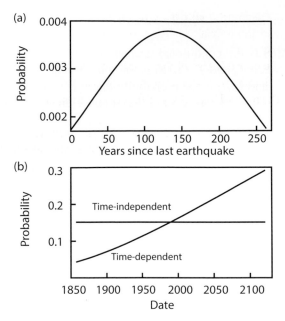

Figure 9.13   (a) Gaussian probability density function for recurrence times. (b) Conditional probability that the earthquake will occur in the next twenty years for two different models.

recurrence time of 132 years, the probability in 20 years is 15%. For times before 1989, the time-independent model predicts higher probabilities. Eventually, at later times, the time-dependent model predicts progressively greater probabilities. This comparison illustrates the *seismic gap* concept: a gap exists when it has been long enough since the last major earthquake that time-dependent models predict an earthquake probability much higher than expected from time-independent models. In this case, the model predicts that the next earthquake is "overdue," and more likely to occur in a gap than elsewhere.

9.5   Shaky Probabilities

As this example for the well-studied San Andreas fault indicates, estimating earthquake probabilities involves deep uncertainty, as discussed in section 4.5. Despite lots of research around the world, we do not know how to best describe them and can have little confidence in the models we use.

To start, we do not know whether to treat earthquake probabilities as time-independent or time-dependent. Both are used, often inconsistently. Often people will show an earthquake hazard map computed with time-independent probabilities, and then speak of an earthquake being "overdue." Obviously, both cannot be correct.

Even once we chose between these options, we can choose between probability density functions. For example, we could use a log-normal distribution, in which the natural logarithm of recurrence time is normally distributed, so recurrence intervals longer than the mean are more likely than shorter ones. Different probability density functions and different parameters for that pdf can be used, because in most cases none fit the actual earthquake history that well.

Although time-dependent models appeal to our seismological instincts, these instincts may be wrong. So far, the data do not show that these models predict the recurrence of future earthquakes significantly better than a time-independent Poisson model. As discussed in section 3.3, the Parkfield earthquake prediction based on a time-dependent model failed, with the earthquake coming long after it was predicted to occur. Moreover, a global test of the seismic gap hypothesis, which examined how well a gap map (Figure 9.14)

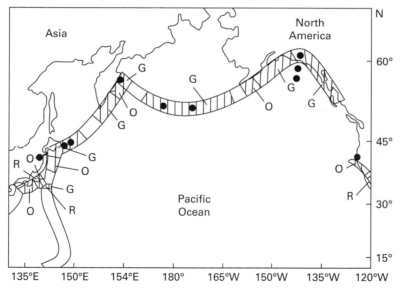

Figure 9.14 Portion of the seismic gap map used by Kagan and Jackson (1991) to test the gap hypothesis. The shaded segments of the plate boundaries had been assigned seismic potentials of high (R), intermediate (O), and low (G). Unshaded segments were regarded as having uncertain potential. During the ten years following the map's publication, ten large (M > 7) earthquakes (dots) occurred in these regions. None were in the high- or intermediate-risk segments, and five were in the low-risk segments. (Stein, 1992. Reproduced with permission of Nature Publishing.)

forecast the locations of major earthquakes, found that the map did no better than random guessing. In fact, many more large earthquakes occurred in areas identified as low risk than in the presumed higher-risk gaps. On some faults, however, time-dependent models seem to work better, so maybe different faults behave differently.

As a result, seismologists are unsure what to think. Our instincts favor the ideas of earthquake cycles and seismic gaps, which let us say more about when earthquakes are due. The Poisson model is less satisfying, but maybe is the best we can do. Our instincts favoring determinism may be incorrect – Einstein initially rejected quantum mechanics, arguing that "God does not play dice."

The key point is that we can estimate earthquake probabilities on a fault in many different ways and get quite different numbers. Hence it does not make sense to talk except in general terms about "the" probability of an earthquake. Freedman and Stark's (2003) view is that "the problem in earthquake forecasts is that the models, unlike the models for coin-tossing, have not been tested against relevant data. Indeed, the models cannot be tested on a human time scale, so there is little reason to believe the probability estimates." Savage (1991) concluded that earthquake probability estimates for California are "virtually meaningless" and that it would be meaningful only to quote broad ranges, such as low ($<10\%$), intermediate ($10–90\%$), or high ($>90\%$).

This issue is a major research challenge. All we can say is that with today's knowledge, any estimated earthquake probability is very uncertain. It is unclear how much this situation will improve as longer paleoseismic histories and other data become available, because they may simply give a better view of how variable earthquake recurrence is in time and space. This uncertainty is a serious problem for hazards policy because estimates of the earthquake hazard and the expected losses over time depend on estimates of earthquake probability, and thus also have large uncertainties.

Questions

9.1. In Figure 9.1, the observed numbers of the largest and smallest earthquakes are less than predicted by the linear relation. Suggest reasons why this occurs.

9.2. The plot in Figure Q9.2 shows the average number of earthquakes per year in the New Madrid seismic zone. About how often do magnitude 5 and 6 earthquakes occur? Use the line extrapolated from the data to estimate about how often magnitude ~7 earthquakes like those in 1811 and 1812 should occur. Another relevant data set is paleoseismic records, which are interpreted as showing that similar large earthquakes occurred in about AD 1450 and AD 900. Estimate a recurrence time from these data and compare it to that implied by the frequency-magnitude plot. What possible interpretations could you draw? Do the GPS data in Figure 4.3 and the earthquake history in Figure 9.12 offer any insight?

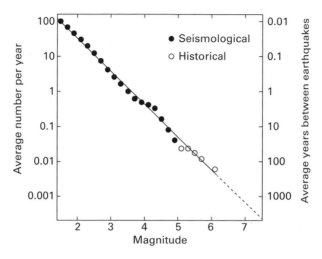

Figure Q9.2   Comparison of earthquakes per year and years between earthquakes. (Richard Aster, Colorado State University. Reproduced with permission.)

9.3.   In the 2011 Tohoku earthquake, an area about 400 km long and 200 km wide slipped about 20 m on average (Figure 9.4). Find what the seismic moment and moment magnitude would have been if the rupture had been 50% longer and the slip had been 50% larger. Assume a rigidity of 3×10^{11} dyne/cm^2 and carefully convert all lengths to cm. Compare the results to Figure 9.4.

9.4.   Given the San Andreas fault data in Figure 9.12 and our discussion of time-independent and time-dependent earthquake probability models, would you say a large earthquake here is overdue? Why or why not? If this record were twice as long but looked similar, how would your view change and why? How would your view have differed if two of the paleoearthquakes in the record had been missed by the paleoseismic studies?

9.5.   As part of its investigation of the 1906 earthquake, members of the commission mulled over issues of earthquake recurrence. G. K. Gilbert posed the crucial question:

   Must the citizens of San Francisco and the bay district face the danger of experiencing within a few generations a shock equal to or even greater

> than the one to which they have just been subjected? Or have they earned
> by their recent calamity a long immunity from violent disturbance? . . .
> If a forecast of immunity shall not be warranted, the public should have the
> benefit of that information, to the end that it shall fully heed the counsel
> of those who maintain that the new city should be earthquake-proof.

Harry Reid estimated that strain released in the earthquake had been accumulating for about a hundred years before 1906, so a similar earthquake would not happen soon. Fusakichi Omori, Japan's leading seismologist, who had joined the commission, agreed that the area "will be free of these great earthquakes for fifty years or more." Rephrase the issue and the arguments in terms of the language used in this chapter, both for science and for policy. Which do you agree and disagree with and why?

9.6. Omori's statement that no earthquakes similar to the 1906 one would occur in the San Francisco area for at least fifty years was based on his experience in Japan, long before the discovery of plate tectonics. Use Figure 9.7 to suggest why his estimate proved correct.

9.7. Write a computer program, or an Excel spreadsheet using the NORMDIST function, that computes the probability of an earthquake using the time-dependent model in equations 9.6, 9.8 and 9.9 (also see problem 4.9). Using the mean and standard deviation in equation 9.10, calculate the probability of a large earthquake between 2013 and 2063 along that segment of the San Andreas. Compare this to that predicted by a time-independent model.

Further Reading and Sources

This discussion follows Stein and Wysession (2003) and Stein (2010). Plate tectonics is discussed in introductory geology texts such as Marshak (2013). Popular books including Hough (2002), Bolt (2006), and Brumbaugh (2010) discuss earthquakes in general and some specifically. Lay and Wallace (1995), Shearer (1999), and Stein and Wysession (2003) discuss seismological topics in more mathematical and physical depth.

Nontechnical online resources include *http://www.seismo.unr.edu/Pre paredness*, *http://www.iris.edu*, *http://www.scec.org/education/*, and *http://earthquake.usgs.gov/learn*.

Geschwind (2001) discusses the 1906 earthquake commission's formation and activities. Its report (Lawson et al., 1908) is available online at *http://content.cdlib.org/view?docId=hb1h4n989f&brand=eqf&chunk.id=meta*.

The Pallett Creek data and analysis are in Sieh et al. (1989). Gilbert's (1906) comments are from an article available at *http://en.wikisource.org/ wiki/Popular_Science_Monthly/Volume_69/August_1906/The_Investigation _of_the_San_Francisco_Earthquake*

References

Bolt, B. A., *Earthquakes*, W. H. Freeman, New York, 2006.

Brumbaugh, D., *Earthquakes: Science and Society*, Prentice Hall, Upper Saddle River, NJ, 2010.

Freedman, D., and P. Stark, What is the chance of an earthquake? in *Earthquake Science and Seismic Risk Reduction*, NATO Science Series IV: Earth and Environmental Sciences, edited by F. Mulargia and R. J. Geller, pp. 201–213, Kluwer, Dordrecht, The Netherlands, 2003.

Geschwind, C.-H., *California Earthquakes: Science, Risk, and the Politics of Hazard Mitigation*, Johns Hopkins University Press, Baltimore, MA, 2001.

Gilbert, G. K., Investigation of the San Francisco earthquake, *Popular Science Monthly*, *69*, August 1906.

Gordon, R. G., and S. Stein, Global tectonics and space geodesy, *Science*, *256*, 333–342, 1992.

Hough, S. E., *Earthshaking Science: What We Know (and Don't Know) about Earthquakes*, Princeton University Press, Princeton, NJ, 2002.

Kagan, Y. Y., and D. D. Jackson, Seismic gap hypothesis: ten years after, *J. Geophys. Res.*, *96*(21), 419–421, 431, 1991.

Lawson, A. C., et al., *The California Earthquake of April 18, 1906: Report of the State Earthquake Investigation Commission*, Carnegie Institution, Washington, DC, 1908.

Lay, T., and T. C. Wallace, *Modern Global Seismology*, Academic Press, New York, 1995.

Marshak, S., *Essentials of Geology*, Norton, New York, 2013.

Savage, J. C., Criticism of some forecasts of the national earthquake prediction council, *Bull. Seismol. Soc. Am.*, *81*, 862–881, 1991.

Shearer, P. M., *Introduction to Seismology*, Cambridge Univ. Press, Cambridge, UK, 1999.

Sieh, K., M. Stuiver, and D. Brillinger, A more precise chronology of earthquakes produced by the San Andreas fault in southern California, *J. Geophys. Res.*, *94*, 603–624, 1989.

Stein, S., Seismic gaps and grizzly bears, *Nature*, *356*, 387–388, 1992.

Stein, S., *Disaster Deferred: How New Science is Changing our View of Earthquake Hazards in the Midwest*, Columbia University Press, New York, 2010.

Stein, S., and M. Wysession, *Introduction to Seismology, Earthquakes, and Earth Structure*, Blackwell, Oxford, 2003.

10

Assessing Hazards

"There are known knowns. These are things we know that we know. There are known unknowns. That is to say, there are things that we know we don't know. But there are also unknown unknowns. There are things we don't know we don't know."

Donald Rumsfeld, US Secretary of Defense

10.1 Five Tough Questions

As the previous chapters illustrate, experience shows that assessing a natural hazard well is challenging. It involves five basic questions:

- *How* is the hazard defined?
- *Where* will major hazardous events occur?
- *When* will they occur?
- *How* big will they be?
- *What* will happen when they occur?

Because our estimates of all the quantities involved have large uncertainties, scientists' goal should be to do the best they can while recognizing the limitations involved. Sometimes this process works well. In other cases it doesn't, indicating the need for improvements. No matter how fancy the hazard assess-

Playing against Nature: Integrating Science and Economics to Mitigate Natural Hazards in an Uncertain World, First Edition. Seth Stein and Jerome Stein.
© 2014 John Wiley & Sons, Ltd. Published 2014 by John Wiley & Sons, Ltd.
Companion Website: www.wiley.com/go/stein/nature

ment algorithm is, or how attractive the brightly colored maps displaying it are, the earth is not obligated to comply and often doesn't.

This chapter illustrates the current status of this process for earthquakes. Earthquake hazard maps often do not do well, largely because earthquakes turn out to be more variable in space, time, and magnitude than implied by the short record available. Although the specifics differ, similar issues arise for other hazards.

10.2   Uncertainties

As we discussed in section 4.5, we can divide uncertainties into shallow and deep uncertainties, depending on how we can treat them mathematically. *Shallow uncertainties* arise when we have reasonable knowledge of the probability distributions (pdfs) of key parameters. For example, in section 8.1 we used historical flood data for a river to estimate how often to expect a flood of a given size. Similarly, in section 8.2 we used a Poisson distribution to estimate how often a hurricane would strike the central Gulf of Mexico coast. In such cases, we can make reasonable estimates of both the average of what we expect and its variability.

In contrast, *deep uncertainties* arise when we lack reasonable knowledge of the probability distributions of key parameters. For example, the statistics of river floods and hurricanes may change significantly in years to come, due to climate change. Because we cannot predict very well how the climate will change and how that change will affect hurricanes or rainfall in a given area, we do not know how the probability distributions used to estimate the recurrence of floods and hurricanes will change. Climate models can give some insight, but these models have their own uncertainties that propagate into the flood and hurricane estimates.

Another example of deep uncertainty arises in trying to estimate earthquake probabilities. As we discussed in section 9.4, these can be estimated in various ways, which give quite different answers. Given our limited understanding of earthquake physics, there is little point in talking about "the" probability of an earthquake – we cannot do any better than to give broad bounds.

It is important to think about which uncertainties can be reduced by more data and by how much. For example, in many places where large earthquakes are rare we do not have good models of the ground shaking that will result from a large earthquake, because we lack the necessary observations. This uncertainty will be reduced eventually when a large earthquake occurs and gives that data. In contrast, because earthquakes are quite variable in space and time, there is probably irreducible uncertainty in trying to estimate earthquake probabilities. Improving our knowledge of earthquake history will help, but even a long and complete history – one that does not miss any large events – probably will still leave considerable uncertainty in what happens next.

Thus when we try to estimate hazards, we want to consider the uncertainties involved, what causes them, and whether these are shallow or deep uncertainties. Doing so helps us to understand how much confidence to place in hazard estimates. We can consider the assumptions made, how well constrained they are by data, and their effects on the estimate. A good way to do this is to look at the estimate's robustness – how does it change as a result of differing assumptions?

10.3   How Is the Hazard Defined?

A natural hazard is not a physical quantity that can be measured. Instead it is a numerical metric chosen for use in mitigation planning and then estimated using a combination of data, the historical record, and models that are assumed to describe aspects of the process in question. As a result, how large a hazard is depends first on how it is defined.

Planning for floods, for example, is based on the water level expected on average at least once in a certain time period, typically 100 or 500 years, or equivalently at a certain probability in a given year. Depending on the application, different measures can be used. Following coastal flooding in 1953 that killed over 1,800 people, the Netherlands has installed systems to protect again the largest flood expected every 10,000 years. The same storm led to the construction of a moveable barrier in the Thames River, designed to protect London from flooding by the largest storm surge expected every 1,000 years.

Analogously, the earthquake hazard in a given location is described by the maximum shaking due to earthquakes that are expected to happen in a given period of time. As we have seen in Chapters 1 and 2, the results are typically presented in hazard maps. The expected shaking is usually given in terms of acceleration, often as fractions of the acceleration of gravity, $g = 9.8\,\mathrm{m/s^2}$, because accelerations are what damage structures. A house on a high-speed train going on a straight track would be unharmed, because the speed is not changing so there is no acceleration. However, if the train stops suddenly, the house will be shaken and could be damaged if the acceleration were large enough.

Hazard maps can be made in various ways, and their relative advantages are debated. The commonly used approach tries to consider all the possible earthquakes that could cause significant shaking at a place. This method, called *probabilistic seismic hazard assessment* or PSHA, involves estimating the probability of different earthquakes and the resulting shaking, and producing an estimate of the combined hazard. Essentially, it is an expected value approach (section 4.1). Despite its widespread use, PSHA yields studies that make tough reading. As people working with it note, "to the benefit of just about no one, its simplicity is deeply veiled by user-hostile notation, antonymous jargon, and proprietary software."

A key feature of this approach is that the estimated hazard depends on the probability or time window, sometimes called *return period*, used. Typically, the probability p that earthquake shaking at a site will exceed some value in next t years, assuming this occurs on average every T years, is assumed to be described by a Poisson distribution (equation 8.18) so

$$p = 1 - e^{-t/\tau} \tag{10.1}$$

which is approximately t/T for $t << T$. Lower probabilities correspond to longer time windows. Thus shaking that there is a 10% chance of exceeding at least once in 50 years will occur on average once about every $50/0.1 = 500$ years (actually 475 years using the more accurate exponential). However shaking with a 2% chance of being exceeded in 50 years will occur on average only every $50/0.02 = 2500$ (actually 2475) years.

The choice of the return period is significant, as illustrated in Figure 10.1 contrasting hazard maps for the US made in 1982 and 1996. The maps show the hazard in terms of peak ground acceleration (PGA). The seismic hazard in the central US New Madrid seismic zone was increased from approximately 1/3 that of the San Andreas fault area in the 1982 map to greater than that of the San Andreas fault in the 1996 map. This change resulted largely from a change in the return period used to define the hazard for building codes. The older map shows the maximum shaking predicted to have a 10% chance of being exceeded at least once in 50 years, which is the criterion used in the maps of most nations. However, the newer map shows the maximum shaking predicted to have a 2% chance of occurring at least once in 50 years.

To see why using a longer return period increases the hazard, consider Figure 10.2, which approximates the New Madrid seismic zone. Earthquakes can be thought of as darts thrown at the map, with the same chance of hitting anywhere in the area shown. About every 150 years a magnitude 6 earthquake hits somewhere, causing moderate shaking in an area that is assumed to be a circle with a radius of 50 km. Over time, more earthquakes hit and a larger portion of the area gets shaken at least once. Some places get shaken a few times. Thus the longer the time period the map covers, the higher the predicted hazard. The new definition in fact increased California's hazard to a level that would have made earthquake resistant construction too expensive, so the hazard there was "capped."

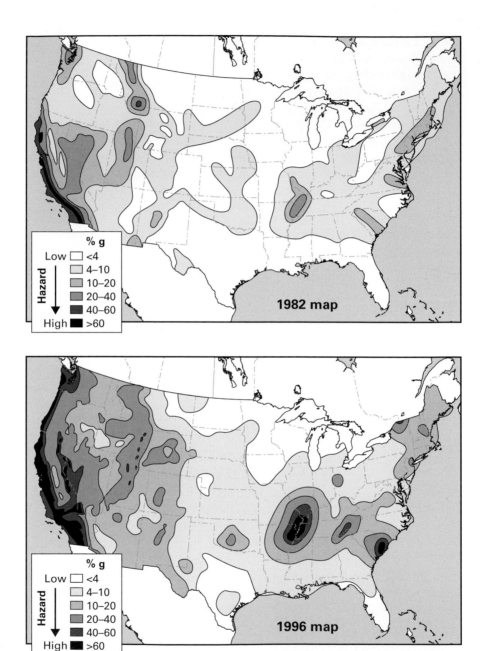

Figure 10.1   Comparison of the 1982 and 1996 US Geological Survey earthquake
hazard maps. The predicted hazard is shown as a percentage of the acceleration of
gravity. Redefining the hazard raised the predicted hazard in the Midwest from being
much less than in California to being even greater than California's. (Stein, 2010.
Reproduced with permission of Columbia University Press.)

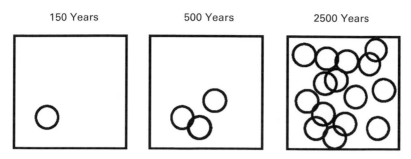

Figure 10.2 Schematic illustration showing how the predicted earthquake hazard increases for longer time windows. The circles show areas within which shaking above a certain level will occur. (Stein, 2010. Reproduced with permission of Columbia University Press.)

This example brings out the point that the definition of hazard involves not only scientific but also political and economic issues. Redefining the hazard using a 2500-year return period rather than 500 years is a decision to require a higher level of seismic safety, via stronger building codes, at higher cost. Whether this makes sense is a complicated and unresolved issue. Typical buildings have a useful life of 50–100 years, so the 2500-year criterion requires that they be designed for shaking that they are unlikely to experience. For comparison, the criterion used in flood planning for normal structures is to plan for the 100-year flood. Different levels may be appropriate for different structures, with extremely stringent hazard requirements for the most critical ones.

An alternative approach is to specify the largest earthquake of concern for each area. That means assuming where it will be, its magnitude, and how much shaking it will cause. This is called *deterministic seismic hazard assessment* or DSHA.

For forty years there has been a debate about which method is better, PSHA or DSHA. The deterministic approach is the simplest, but makes society spend lots of money preparing for an earthquake that is very unlikely to happen during a structure's life. Its advocates especially favor it for a critical facility like a nuclear power plant, whose failure would be so serious that it is worth spending a lot to avoid. The probabilistic approach is closer to how people decide whether to buy insurance or safety devices, because they consider how likely something is to happen when deciding how much to invest in protecting themselves. Critics of PSHA dislike defining the hazard in terms of a mathematical event rather than a real earthquake. They point out that assuming that quantities like ground shaking are described by probability density functions allows for extreme values, at low probabilities, that are physically implausible.

In some cases, the choice of method makes a big difference, and in others it does not. Both are used, and some approaches combine the two. As probabilistic models cover longer time windows they become about the same as deterministic ones. As Figure 10.2 shows, after a long enough time any earthquake that is going to happen will have happened at least once. Moreover, many of the problems resulting from limited knowledge, such as our lack of knowledge of how big the biggest earthquake in an area can be, apply to both methods.

10.4   Where Will Large Earthquakes Occur?

Once a hazard mapper decides how to define the hazard, the remaining questions are geological rather than political. The next question is where to assume that earthquakes will happen.

The simplest – but not always best – assumption is that earthquakes will happen where they are known to have happened. On plate boundaries, that is a good if not perfect assumption. As we discussed for Japan in section 2.1 and the San Andreas fault in section 9.3, plate motion causes strain to accumulate, which we can see occurring with GPS data. Thus the earthquake history on a plate boundary developed from earthquakes that were either seismologically recorded, known from historical records, or shown by paleoseismology is a pretty good guide to where to expect large earthquakes. Which pieces or segments will break next, and in what combination, is harder to predict because the space-time patterns vary (Figure 2.4).

The faster the motion on the boundary, the more often large earthquakes release the accumulated strain, so on some parts of slowly moving boundaries we may not have a record of large earthquakes. However, we expect that earthquakes will eventually happen there. Figure 10.3 illustrates this for the coast of North Africa, which is part of the slow-moving convergent plate boundary between west Africa (called Nubia) and Eurasia. During the time period over which we have good seismological data, roughly since 1910, only parts of the boundary have had large (magnitude 7) earthquakes. However, if we assume that these occur at about the rate they have been, about one every 50 years, a simple numerical simulation implies that over thousands of years the whole boundary will have them.

This simulation brings out the danger that a too-short earthquake history will lead us to focus on apparent patterns that that result from the short history. For example, imagine that we had 2,000 years of data – a longer record than available in most places – that were complete in that no major earthquakes were missed. Even so, the fact that some parts of the boundary look active and others do not would just be an artifact, and we would need 8,000 years

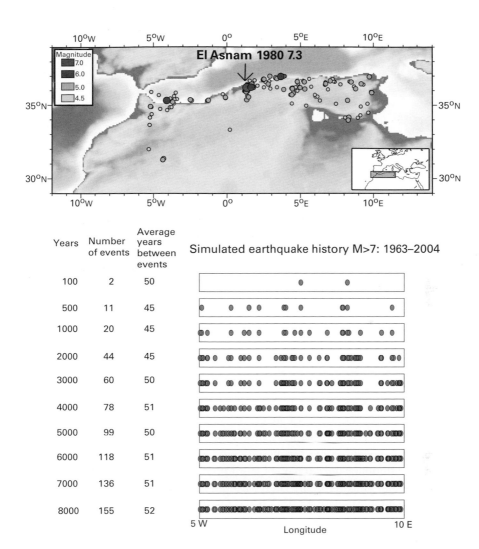

Figure 10.3 Seismicity along the North Africa plate boundary for 1963–2004. Simulations using a frequency-magnitude relation derived from these data predict that if seismicity is uniform in the zone, about an 8000-year record is needed to avoid apparent concentrations and gaps. (Swafford and Stein, 2007. Reproduced with permission of the Geological Society of America.) See also color plate 10.3.

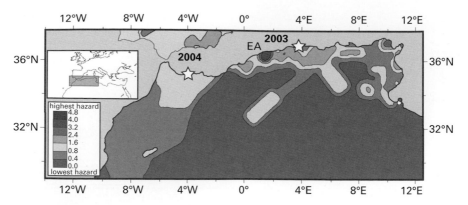

Figure 10.4   Global Seismic Hazard Map (1999) for North Africa, showing peak ground acceleration in m/s² expected at 10% probability in 50 years. Note "bull's-eye" at site of the 1980 M_s 7.3 El Asnam (EA) earthquake. The largest subsequent earthquakes to date, the May 2003 M_s 6.8 Algeria and February 2004 M_s 6.4 Morocco events (stars) did not occur in the predicted high hazard regions. (Swafford and Stein, 2007. Reproduced with permission of the Geological Society of America.)

to see that the seismicity was the same along the boundary. Because in most areas our record is much shorter, it would be easy to reach an incorrect conclusion. This seems to have happened, in that the earthquake hazard map for the area shows high-hazard bull's-eyes where known past earthquakes occurred, but the largest earthquakes since the map was made happened in places mapped as low hazard (Figure 10.4).

Similarly, because only seismologically recorded earthquakes were used, the 2001 earthquake hazard map of Haiti predicted low hazard in the area that suffered major shaking in the 2010 earthquake (Figures 1.4, 1.5). A different picture would have emerged from using the record of historic earthquakes on known faults and GPS data showing strain accumulating on these faults.

Another complication is that the faults in Haiti are part of the boundary between the North American and Caribbean plates, which is a broad zone rather than a narrow boundary. The plate motion is spread out over a number of faults, so knowing the overall plate motion across the zone does not tell us how it is distributed and thus what to expect in terms of earthquakes. However, geologic data can be used to study the past motion on individual faults and GPS data can show the present strain accumulation on each.

Many plate boundaries are actually boundary zones. For example, although most of the motion between the Pacific and North American plates in the US occurs along the San Andreas fault, some is taken up by other faults in a broad zone extending eastward across Nevada and Utah, on which large earthquakes occur. Similarly, the 2008 Wenchuan earthquake (M 7.9) in Sichuan Province, China, occurred on the Longmenshan fault, which is part of the broad boundary between the Indian and Eurasian plates (Figure 9.7). This fault was assessed, based on the lack of recent seismicity (Figure 10.5) to have low hazard. However, a different view would have come from considering the geology and GPS data that show 1–3 mm/yr of motion across the Longmenshan Fault.

Although this seems slow, over 500–1000 years enough motion would accumulate for a magnitude 7 earthquake, and longer intervals would permit even larger earthquakes.

Inferring where earthquakes will happen can get even trickier within plates. In some places within continents, large earthquakes cluster on specific faults for some time and then migrate to others (Figure 10.6). These faults "turn on" and "turn off" on timescales of hundreds or thousands of years, causing episodic, clustered, and migrating large earthquakes, which make it hard to assess the extent to which the seismological record reflects the location of future earthquakes. If earthquakes migrate, predicting where big earthquakes will happen is like the carnival game "Whack-a-mole." You usually won't hit the mole by waiting for it to come up where it went down, because it will pop up somewhere else.

A striking example is in North China (Figure 10.7), where a 2000-year record shows migration of large earthquakes between fault systems spread over a broad region, such that no large earthquake ruptured the same fault segment twice in this interval. Making a map using any short subset of the record would be biased. For example, a map using the 1900 years prior to 1950 would miss the recent activity in the North China plain, including the 1976 Tangshan earthquake (M_w 7.8), which occurred on a previously unknown fault and killed nearly 240,000 people. Earthquakes on faults that had not been previously identified, often because they have no clear surface expression, also occur in many other areas.

To make things even more complicated, many large earthquakes within continents have aftershock sequences that last hundreds of years or even longer. As a result, many small earthquakes may be aftershocks of previous large quakes. In such cases, treating small earthquakes as indicating the location of future large ones can overestimate the hazard in presently active areas and underestimate it elsewhere.

Thus within plates, where to assume large earthquakes will happen is crucial in hazard mapping. Figure 10.8 illustrates this issue by comparing hazard maps for Canada made in 1985 and 2005. The older map shows concentrated high hazard bull's-eyes along the east coast at the sites of the 1929 M 7.3 Grand Banks and 1933 M 7.4 Baffin Bay earthquakes, on the

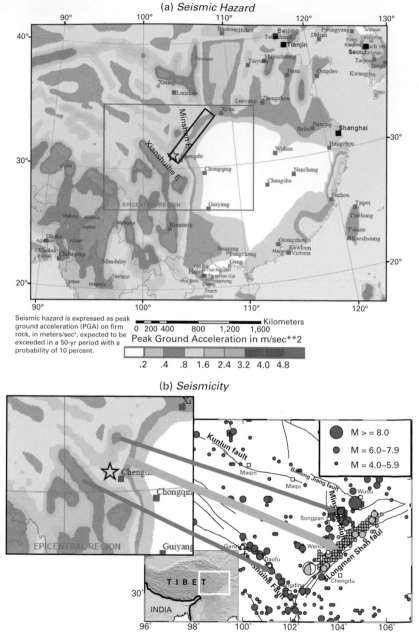

Figure 10.5 (a) Seismic hazard map for China produced prior to the 2008 Wenchuan earthquake, which occurred on the Longmenshan Fault (black rectangle). (b) Seismicity in the region. The hazard map showed low hazard on the Longmenshan fault, on which little instrumentally recorded seismicity had occurred before the Wenchuan earthquake, and higher hazard on faults nearby that showed more seismicity. (Stein et al., 2012. Reproduced with permission of Elsevier, B.V.) See also color plate 10.5.

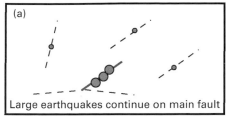
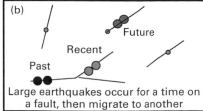

Figure 10.6   Episodic, clustered, and migrating large earthquakes. In many continental fault systems, it appears that rather than one main fault staying active for a long time (a), many faults turn on and off (b). (McKenna et al., 2007. Reproduced with permission of the Geological Society of America.)

assumption that there is something especially hazardous about these locations. The alternative is to assume that similar earthquakes can occur anywhere along the margin, because the entire margin contains faults remaining from the continental rifting that opened up the Atlantic Ocean. The 2005 map makes this assumption, and thus shows a "ribbon" of high hazard along the coast, while still retaining the bull's-eyes. This demonstrates that quite different maps result depending on whether only the instrumentally recorded earthquakes data is used, or whether geological assumptions are included.

10.5   When Will Large Earthquakes Occur?

Hazard mapping requires assuming some probability density function that describes the distribution of future earthquake recurrence intervals, so as to predict the expected shaking in a time period. As we have discussed in previous chapters, it is unclear what to assume, given that there are many choices and in most cases none do a good job of describing the complicated earthquake history.

As discussed in Chapter 9, a major choice for mapping is between time-independent models, in which a future large earthquake is equally likely immediately after the past one and much later, and time-dependent models, in which the probability is small shortly after the past one, and then increases with time. The latter involves choosing one of the many probability density functions available, and both involve estimating parameters using the limited earthquake record.

The effect on hazard models depends on the assumed recurrence interval and how long it has been since the past large earthquake. For times since the previous earthquake that are only a small fraction of the assumed mean

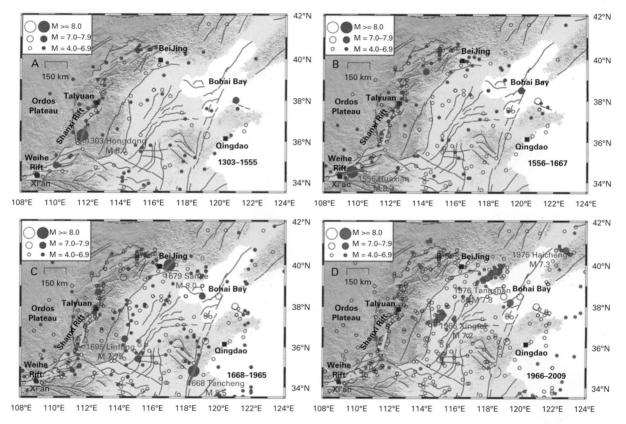

Figure 10.7 Earthquake history of North China. Solid circles are locations of events during the period shown in each panel; open circles are locations of events from 780 BC to the end of the previous period (1303 AD for panel A). Bars show the rupture lengths for selected large events. (Liu et al., 2011. Reproduced with permission of the Geological Society of America.) See also color plate 10.7.

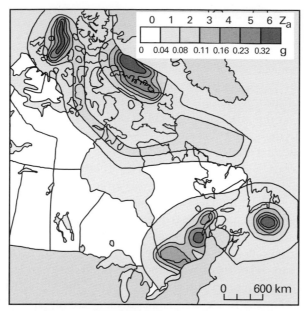

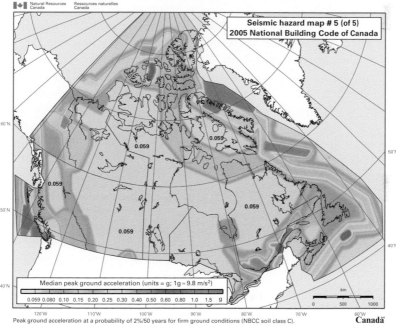

Figure 10.8 Comparison of the 1985 and 2005 Geological Survey of Canada earthquake hazard maps of Canada. The older map shows concentrated high hazard bull's-eyes along the east coast at the sites of the 1929 Grand Banks and 1933 Baffin Bay earthquakes, whereas the new map assumes that similar earthquakes can occur anywhere along the margin. (Stein et al., 2012. Reproduced with permission of Elsevier, B.V.) See also color plate 10.8.

recurrence interval, time-dependent models predict lower probabilities. Eventually, if a large earthquake has not occurred by this time, the earthquake is "overdue" in the sense that time-dependent models predict higher probabilities. Figure 10.9 (top) illustrates this effect for faults on which the time since the past large earthquake is a different fraction of the assumed mean recurrence interval.

The difference is shown in Figure 10.9 (bottom) for the New Madrid seismic zone, where M7 earthquakes occurred in the period 1811–1812. Assuming that such earthquakes occur on average every 500 years, the probability of having one in the next 50 years is 50/500 or 10% in a time-independent model. Alternative models assume that recurrence times have Gaussian distributions with a mean of 500 years and a standard deviation of 100 or 200 years. The "crossover" time is around the year 2144, 333 years after 1811. The predictions of the time-dependent models of the probability of a major earthquake in the next hundred years are much smaller than that of the time-independent model. Similar results arise for a time-dependent model with recurrence times described by a lognormal probability distribution.

The effect of the model choice on a hazard map is illustrated in Figure 10.10 by alternative maps for the New Madrid zone. The biggest effect is close to the three faults used to model the jagged geometry of the earthquakes of 1811–1812, where the largest hazard is predicted. Compared to the hazard predicted by the time-independent model, the time-dependent model predicts noticeably lower hazard for the 50-year periods 2000–2050 and 2100–2150. For example in Memphis, the time-dependent model predicts hazards for 2000–2050 and 2100–2150 that are 64% and 84% of those predicted by the time-independent model. However if a large earthquake has not occurred by 2200, the hazard predicted in the next 50 years would be higher than predicted by the time-independent model.

Given the limitations of knowledge, choosing how to model the recurrence on faults in a hazard map largely reflects the mappers' preconceptions. Thus the Japanese map (Figure 2.1) made prior to the Tohoku earthquake reflected the mappers' view that a large earthquake would happen much sooner on the Nankai Trough than off Tohoku.

10.6   How Big Will the Large Earthquakes Be?

Even after the location and recurrence time of major earthquakes are assumed, hazard maps also depend dramatically on the assumed magnitude of the largest earthquakes expected in each area, denoted as M_{max}. This crucial quantity is unknown and not at present effectively predictable on physical

(a) Conditional probability of earthquake in next t years

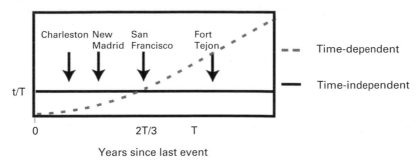

Years since last event

(b) Conditional probabilities of earthquake in next 50 years

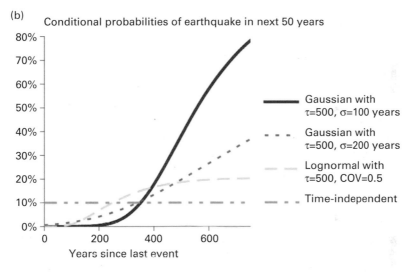

Years since last event

Figure 10.9 (a) Schematic comparison of time-independent and time-dependent models for different seismic zones. Charleston and New Madrid are "early" in their cycles, so time-dependent models predict lower hazards. The two model types predict essentially the same hazard for a recurrence of the 1906 San Francisco earthquake, and time-dependent models predict higher hazard for the nominally "overdue" recurrence of the 1857 Fort Tejon earthquake. The time-dependent curve is schematic because its shape depends on the probability distribution and its parameters. (b) Comparison of the conditional probability of a large earthquake in the New Madrid zone in the next 50 years, assuming that the mean recurrence time is 500 years. In the time-independent model the probability is 10%. Time-dependent models predict lower probabilities of a large earthquake for the next hundred years. (Hebden and Stein, 2009. Reproduced with permission of the Seismological Society of America.) See also color plate 10.9.

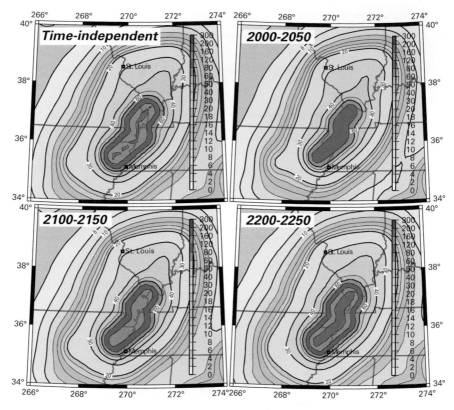

Figure 10.10   Comparison of hazard maps for the New Madrid zone. Shading shows peak ground acceleration as percentages of 1 g. Compared to the hazard predicted by the time-independent model, the time-dependent model predicts noticeably lower hazard for the periods 2000–2050 and 2100–2150, but higher hazard if a large earthquake has not occurred by 2200. (Stein et al., 2012. Reproduced with permission of Elsevier, B.V.) See also color plate 10.10.

grounds, because even where we know the long-term rate of motion across a plate boundary fault, or the deformation rate across an intraplate zone, neither predict how strain will be released. As a result, quite different M_{max} estimates can be made. Mappers predict the size and rate of future large earthquakes from an earthquake record containing seismological, geological, and historical earthquake records. The Tohoku, Wenchuan, and Haiti examples, in which the earthquake was much larger than the M_{max} assumed, indicate that this process is far from straightforward and prone to a variety of biases. In particular, because catalogs are often short relative to the average recurrence time of large earthquakes, larger earthquakes than anticipated often occur.

As we discussed in section 9.2, the Gutenberg–Richter frequency-magnitude relation lets us combine seismological data for smaller earthquakes with paleoseismic data or geologic inferences for larger earthquakes about which we lack adequate data, due to the limited earthquake history. However, often these do not match well – large earthquakes occur more or less frequently than expected from the log-linear relation observed for smaller earthquakes. Moreover, we have no way of knowing whether the largest earthquake we know of for some area is really the largest that happens there, or just the largest that we have observed or know of. As the 2011 Tohoku and 2004 Sumatra earthquakes showed, often a much bigger earthquake than expected occurs.

One way to explore this question is with numerical simulations like that in Figure 10.11, which involved generating synthetic earthquake histories of various lengths assuming that the seismicity followed a Gutenberg–Richter

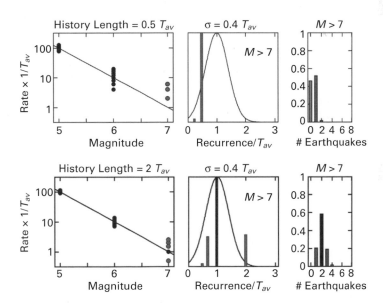

Figure 10.11 Results of numerical simulations of earthquake sequences. Rows show results for sequences of different lengths. Left panels show the log-linear frequency-magnitude relation sampled, with dots showing the resulting mean recurrence times. Center panels show the parent distribution of recurrence times for $M \geq 7$ earthquakes (smooth curve) and the observed mean recurrence times (bars). Right panels show the fraction of sequences in which a given number of $M \geq 7$ earthquakes occurred. (Stein and Newman, 2004. Reproduced with permission of the Seismological Society of America.) See also color plate 10.11.

relation. The recurrence times of earthquakes with M $\geq$ 5, 6, and 7 were assumed to be samples of a Gaussian (normal) parent distribution with a standard deviation of 0.4 times the mean recurrence for each of the three magnitudes.

Figure 10.11 (top) shows the results for 10,000 synthetic earthquake sequences whose length is half of T_{av}, the mean recurrence time of earthquakes with M $\geq$ 7. The left panel has a line showing the log-linear frequency-magnitude relation that was sampled and dots marking the "observed" mean recurrence rates in the simulation for M $\geq$ 5, 6, and 7 events in each sequence. The center panel shows the parent distribution of recurrence times for M $\geq$ 7 that was sampled, and a histogram of the "observed" mean recurrence times for the sequences. Due to their short length, 46% of the sequences contain no earthquakes with M $\geq$ 7. In these cases, we would underestimate the size of the largest earthquakes that can occur and thus underestimate the hazard. In the other cases, we are most likely to "observe" large earthquakes whose recurrence time is less than the average, so we overestimate the rate of the largest earthquakes, and thus also overestimate the seismic hazard. For longer sequences we are likely to observe the largest earthquakes, but our estimate of their rate can be poor. As shown in the bottom panels, even if the sequence length is twice T_{av}, we underestimate the rate of the largest earthquakes 20% of the time and overestimate it another 20% of the time.

Other biases can arise from the challenge of estimating the magnitude and rate of historic earthquakes and paleoearthquakes. If the magnitudes of historic earthquakes or paleoearthquakes were overestimated, a Gutenberg–Richter curve would lead us to think they happen more frequently than they do. Conversely, the mean recurrence time would be overestimated if some paleoearthquakes in a series were missed.

10.7   How Much Shaking?

Hazard mapping also has to assume how much shaking future earthquakes will produce. This involves adopting a *ground motion attenuation relation*, which predicts the ground motion expected at a given distance from earthquakes of a given size. Different relations predict quite different hazards.

Figure 10.12 illustrates this by comparing various ground motion models for the central US. In general, the predicted shaking decreases rapidly with distance for a given magnitude – note the logarithmic scale. Thus although shaking is felt over large distances, serious damage is much more concentrated. A striking example is the fact that much less damage resulted from shaking in the giant Tohoku and Sumatra earthquakes than from the tsunamis,

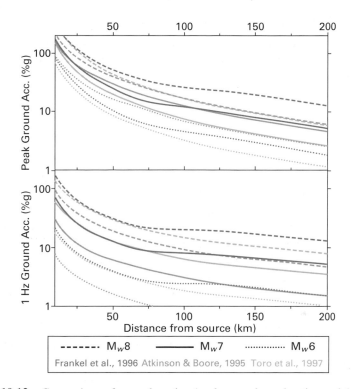

Figure 10.12   Comparison of ground motion (peak ground acceleration and 1 Hz) as a function of distance for three different earthquake magnitudes predicted by three models for the central US. For $M_w 8$, the Frankel et al. (1996) model predicts significantly higher values than the others. (Newman et al., 2001. Reproduced with permission of the Seismological Society of America.) See also color plate 10.12.

because the earthquakes occurred offshore. The models also show that a smaller earthquake nearby can cause more shaking and thus damage than a larger one farther away.

In areas that are seismically active enough, seismological observations can be used to develop ground motion prediction models describing average ground motion as a function of distance, about which actual motions scatter owing to variations in crustal structure and earthquake source properties. As additional data accumulate, these models get better. However, in many areas, including the central US, there are no seismological records of shaking from large earthquakes. In such cases, mappers choose between various relations derived using data from smaller earthquakes and earthquake source models, which can predict quite different ground motion and thus hazard.

For example, in Figure 10.12 the model labeled "Frankel" predicts significantly higher ground motion than the other two models. In fact, the ground motion for a magnitude 7 earthquake predicted by the Frankel model at distances greater than 100 km is comparable to that predicted for an magnitude 8 earthquake by the other models.

The assumed maximum magnitude and assumed ground motion model have related and offsetting effects in hazard maps, because a similar shaking pattern can result from a smaller earthquake and a model that predicts more shaking for a given magnitude. The effect of these choices is shown in Figure 10.13 by four maps. In each row are two maps for the same ground motion model, but different values of the maximum magnitude – the magnitude of the largest earthquake on the main faults. Raising this magnitude from 7 to 8 increases the predicted hazard at St. Louis by about 35%. For Memphis, which is closer to the main faults, the increase is even greater. This is because the assumed maximum magnitude of the largest earthquake on the main faults affects the predicted hazard, especially near those faults.

The two maps in each column have the same maximum magnitude but different ground motion models. The "Frankel" model predicts a hazard in St. Louis about 80% higher than that predicted by the "Toro" model. For Memphis, this increase is about 30%. The ground motion model affects the predicted hazard all over the area, because shaking results both from the largest earthquakes and from smaller earthquakes off the main faults.

10.8   Dealing With the Uncertainties

Hazard maps clearly have large uncertainties. When a map fails, it is often clear in hindsight that key parameters were poorly estimated. The examples we have looked at show the same point – the maps are uncertain in the sense that their predictions vary significantly depending on the choice of many poorly known parameters.

A way to see this is to compare hazard maps. Figure 10.14 (a copy of Figure 5.7 reproduced here for convenience) compares the predictions of the models in Figures 10.10 and 10.13 for the hazard at St. Louis and Memphis. The predictions vary by a factor of more than three. This representation shows the effects of the three factors. At Memphis, close to the main faults, the primary effect is that of magnitude, with the two magnitude 8 models predicting the highest hazard. At St. Louis, the ground motion model has the largest effect, so the "Frankel" models predict the highest hazard. In fact, the uncertainty is even bigger than these six maps show, because the effect of choosing between time-independent and time-dependent models is shown for

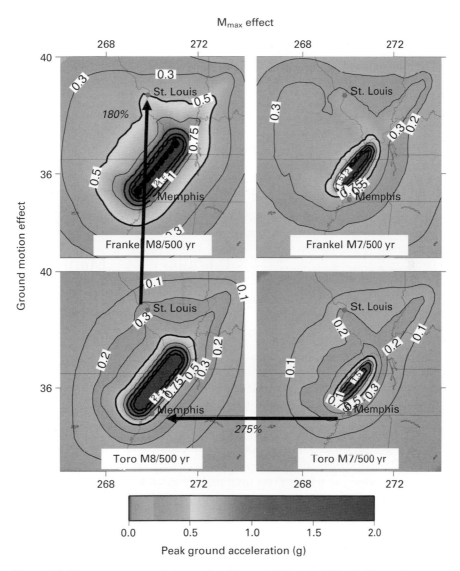

Figure 10.13   Comparison of the predicted hazard (2% probability in 50 years) showing the effect of different ground motion models and maximum magnitudes of the New Madrid fault source. (Newman et al., 2001. Reproduced with permission of the Seismological Society of America.) See also color plate 10.13.

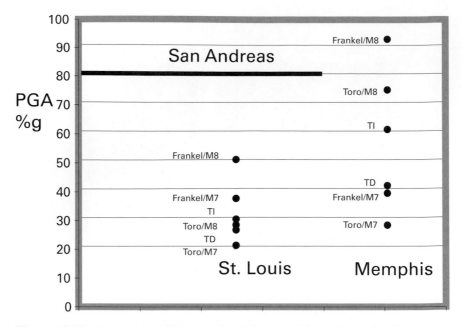

Figure 10.14 Comparison of the hazard at St Louis and Memphis predicted by the different hazard maps of the New Madrid zone shown in Figures 10.10 and 10.13. For example, Frankel/M8 indicates the Frankel et al. (1996) ground motion model with a maximum magnitude of 8 in Figure 10.13, and TI indicates the time-independent model in Figure 10.10. (Stein et al., 2012. Reproduced with permission of Elsevier, B.V.)

a specific set of parameters and a specific combination of maximum magnitude and ground motion model. In reality, for each combination of maximum magnitude and ground motion model, there would be a range of predicted hazard depending on whether one used time-independent and time-dependent probability models, and the parameters used in each.

All of these uncertainties are deep uncertainties. Even if we knew how large the earthquakes of 1811 and 1812 had been – the best present estimates are about magnitude 7 or slightly lower – we would not know how large future ones would be. We do not know whether to treat earthquake recurrence as time-independent or time-dependent even on plate boundaries, so we certainly do not within plates where large earthquakes are much more variable in space and time. Until a large earthquake happens, we will not know how to describe the expected ground motion well. As a result, these uncertainties cannot be reduced by any knowledge we are likely to acquire soon, or be adequately described by a probability density function. The best we can do is give a general sense of about how large they are likely to be.

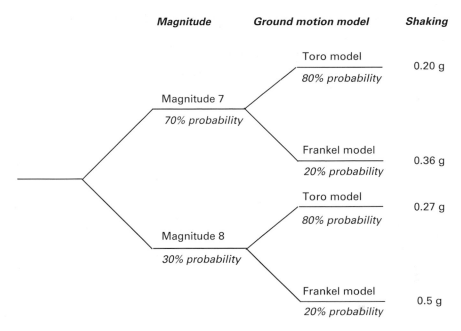

Figure 10.15 Logic tree combining the four models in Figure 10.13 to predict shaking at St. Louis. (Stein, 2010. Reproduced with permission of Columbia University Press.)

Unfortunately, uncertainties like these are not usually communicated to users of hazard maps. Typically, mappers combine the predictions for various parameters via a "logic tree" in which they assign weights that they think represent their relative probabilities. For example, Figure 10.15 shows how the four models in Figure 10.13 can be combined. Starting on the left, there are two different magnitudes for the largest earthquake. For each, there are two different ground motion models. This gives four "branches," of the tree. Each branch has a different prediction for the level of shaking at any point on the map, as shown by the values for St. Louis.

These predictions are combined by assigning weights to each branch. The values shown assign a 70% probability to magnitude 7, and 30% to magnitude 8, which is similar to saying that the magnitude is $(0.70 \times 7) + (0.30 \times 8) = 7.3$. The values shown assign 80% probability to the Toro ground motion model and 20% to the Frankel model. The numbers on each branch are then multiplied. For example, the top branch gives $0.70 \times 0.80 \times 0.20 = 0.11$ g, or 11% of the acceleration of gravity. Adding the four branches gives a value of 0.26 g, or 26% of the acceleration of gravity. This is the predicted value for the shaking at St. Louis coming from a weighted combination of the four different predictions. This expected value is somewhere between the four predictions, and is most influenced by the branches to which the highest probability was assigned.

Adjusting the weights changes the predicted hazard. Because there is no objective way to assign the weights, the result – which often will not be known for hundreds of years or longer – will be as good or as bad as the preconceptions the hazard mappers used to assign the weights ultimately turn out to be. As we have seen, sometimes these prove to have been poor choices. Moreover, showing the resulting single map does not convey the uncertainty involved.

Instead, it is important that uncertainties in hazard map predictions should be assessed, which is hard. As we have seen in section 5.1, in many applications uncertainties turn out to be bigger than had been thought. Reasonable estimates of these uncertainties need to be clearly communicated to potential users. Recognizing the uncertainties – even if they are not that well known and probably underestimated – would help users to decide how much credence to place in the maps and make them more useful in formulating cost-effective hazard mitigation policies, as we will discuss in Chapter 11.

10.9   Next Steps

In an ideal world, these issues would have been worked out before communities started using hazard maps to make multi-billion dollar decisions. Once we knew how well the approach forecast future hazards, we could have chosen whether and how to use it with full awareness of its limitations while trying to improve it.

However, humans being what we are, our instinct, as in medicine (section 5.3) or other areas, is to try something and see how it works. As we have discussed, this sometimes leads to embarrassing results.

Beyond figuring out how to assess and communicate the uncertainties of current maps, the question is what to do to improve them. One obvious approach is to improve the science going into them. As we learn more about earthquakes in specific areas and earthquake physics, we expect that this knowledge will improve forecasts, although there are probably limits as to how well we can ever do.

The other obvious approach is to develop methods for testing hazard maps, as discussed in section 5.3. Testing will show how well maps actually work at present, whether different approaches would work better, and how much new data and methods improve maps' performance.

Questions

10.1.   Earthquake hazard maps typically show high-hazard bull's-eyes at the locations of past large earthquakes. Are such maps being made with time-independent or time-dependent probabilities?

10.2.   In making a hazard map of a geologically similar region, like the plate boundaries along Japan Trench or North African coast, one can assume that:

a)   Areas within the region where no large earthquakes occurred in the past few hundred years are less likely to have them in the next 50–100 years and are thus less hazardous.

b)   Areas within the region where no large earthquakes occurred in the past few hundred years are "gaps" likely to have them in the next 50–100 years and are thus more hazardous.

c)   All areas within the region are equally likely to have large earthquakes and are thus equally hazardous.

Which assumption would you favor and why?

10.3.   From the data shown in Figure 10.9, would you favor stronger and more expensive building standards near the "Fort Tejon" segment of the San Andreas fault than in San Francisco? Why or why not? Conversely, would you favor less stringent and less expensive standards after a large earthquake? Why or why not?

10.4.   On physical grounds, why would you expect the shaking from an earthquake to decrease so rapidly with distance from the earthquake?

10.5.   Using the results shown in Figure 10.14, assess what fraction of the uncertainty in the estimated hazard results from the uncertainty in the ground motion model.

10.6.   From the results shown in Figure 10.14, how would you compare the earthquake hazard in St. Louis and Memphis to that in California?

10.7.   Write a program or spreadsheet to calculate the logic tree in Figure 10.15. How could you change the weights on the magnitude and ground motion model branches if you wanted to raise the hazard from $0.26\,g$ to $0.4\,g$? How could you lower it to $0.22\,g$?

Further Reading and Sources

The Rumsfeld quotation is from a Defense Department briefing on February 12, 2002. The "deeply veiled" quotation is from Hanks and Cornell (1994).

This discussion of earthquake hazard mapping follows Stein (2009, 2010) and Stein et al. (2011, 2012). The latter is a review paper with references to the various topics involved. Castanos and Lomnitz (2002), Mucciarelli et al. (2008), Bommer (2009), Panza et al. (2010), Wang (2011), Peresan and Panza (2012), Wang and Cobb (2012), Baker (2013), Pease (2013), and Gulkan (2013) discuss earthquake hazard assessment methodology.

Searer et al. (2007) discuss the 2500-year criterion. Web resources about earthquake hazard mapping include *http://www.globalquakemodel.org*, *http://earthquake.usgs.gov/hazards/*, *http://www.earthquakescanada.nrcan.gc.ca/hazard/*, *http://www.j-shis.bosai.go.jp/en/shm* and *http://www.efehr.org:8080/jetspeed/portal/*.

References

Baker, J., Quake catcher, *Nature*, *498*, 290–292, 2013.

Bommer, J. J., Deterministic vs. probabilistic seismic hazard assessment: an exaggerated and obstructive dichotomy, *J. Earthq. Eng.*, *6*, 43–73, 2009.

Castanos, H., and C. Lomnitz, PSHA: is it science? *Eng. Geol.*, *66*, 315–317, 2002.

Frankel, A., C. Mueller, T. Barnhard, D. Perkins, E. Leyendecker, N. Dickman, S. Harmson, and M. Hopper, *National Seismic Hazard Maps Documentation*, U.S. Geological Survey Open-File Report 96–532, 1996.

Gulkan, P., A dispassionate view of seismic-hazard assessment, *Seismol. Res. Lett.*, *84*, 413–416, 2013.

Hanks, T. C., and C. A. Cornell, Probabilistic seismic hazard analysis: a beginner's guide, *Fifth Symposium on Current Issues Related to Nuclear Power Plant Structures, Equipment, and Piping*, 1994.

Hebden, J., and S. Stein, Time-dependent seismic hazard maps for the New Madrid seismic zone and Charleston, South Carolina areas, *Seismol. Res. Lett.*, *80*, 10–20, 2009.

Liu, M., S. Stein, and H. Wang, 2000 years of migrating earthquakes in North China: how earthquakes in mid-continents differ from those at plate boundaries, *Lithosphere*, *3*, 128–132, doi: 10.1130/L129, 2011.

McKenna, J., S. Stein, and C. A. Stein, Is the New Madrid seismic zone hotter and weaker than its surroundings?, in *Continental Intraplate Earthquakes: Science, Hazard, and Policy Issues*, Special Paper 425, edited by S. Stein and S. Mazzotti, pp. 167–175, Geol. Soc. Amer., Boulder, CO, 2007.

Mucciarelli, M., D. Albarello, and V. D'Amico, Comparison of probabilistic seismic hazard estimates in Italy, *Bull. Seismol. Soc. Am.*, *98*, 2652–2664, 2008.

Newman, A., S. Stein, J. Schneider, and A. Mendez, Uncertainties in seismic hazard maps for the New Madrid Seismic Zone, *Seismol. Res. Lett.*, *72*, 653–667, 2001.

Panza, G. F., K. Irikura, M. Kouteva, A. Peresan, Z. Wang, and R. Saragoni, Advanced seismic hazard assessment, *Pure Appl. Geophys.*, *168*, 1–9, 2010.

Pease, R., Seismic data set could improve earthquake forecasting, http://news.sciencemag.org/sciencenow/2013/06/seismic-data-set-could-improve-e.html, 2013.

Peresan, A., and G. Panza, Improving earthquake hazard assessments in Italy: an alternative to Texas sharpshooting, *Eos Trans. AGU*, *93*, 538, 2012.

Searer, G., S. A. Freeman, and T. F. Paret, Does it make sense from engineering and scientific perspectives to design for a 2475-year earthquake?, in *Continental Intraplate Earthquakes, Science, Hazard, and Policy Issues*, Special Paper 425, edited by S. Stein and S. Mazzotti, pp. 353–361, Geol. Soc. Amer., Boulder, CO, 2007.

Stein, S., Understanding earthquake hazard maps, *Earth*, January, 24–31, 2009.

Stein, S., *Disaster Deferred: How New Science is Changing our View of Earthquake Hazards in the Midwest*, Columbia University Press, New York, 2010.

Stein, S., and A. Newman, Characteristic and uncharacteristic earthquakes as possible artifacts: applications to the New Madrid and Wabash seismic zones, *Seismol. Res. Lett.*, *75*, 170–184, 2004.

Stein, S., R. J. Geller, and M. Liu, Bad assumptions or bad luck: why earthquake hazard maps need objective testing, *Seismol. Res. Lett.*, *82*, 623–626, 2011.

Stein, S., R. J. Geller, and M. Liu, Why earthquake hazard maps often fail and what to do about it, *Tectonophysics*, *562–563*, 623–626, 2012.

Swafford, L., and S. Stein, Limitations of the short earthquake record for seismicity and seismic hazard studies, in *Continental Intraplate Earthquakes, Science, Hazard, and Policy Issues*, Special Paper 425, edited by S. Stein and S. Mazzotti, pp. 49–58, Geol. Soc. Amer., Boulder, CO, 2007.

Wang, Z., Seismic hazard assessment: issues and alternatives, *Pure Appl. Geophys.*, *168*, 11–25, 2011.

Wang, Z., and J. Cobb, A critique of probabilistic versus deterministic seismic hazard assessment analysis with special reference to the New Madrid seismic zone, in *Recent Advances in North American Paleoseismology and Neotectonics East of the Rockies*, Special Paper 493, edited by R. Cox, M. Tuttle, O. Boyd, and J. Locat, pp. 259–275, Geol. Soc. Amer., Boulder, CO, 2012.

11

Mitigating Hazards

"I was learning that choices in war are rarely between good and bad, but between bad and worse."

Nathan Fick, *One Bullet Away: The Making of a Marine Officer, 2006*[1]

11.1 Approaches

The goal of assessing natural hazards is to develop strategies to cope with them. This task can be viewed as part of the discipline called risk management, which explores how individuals, organizations, and communities can identify, assess, and minimize the effects of undesirable events. These events could be natural, financial, legal or political. Many different terms are used for similar concepts. In this book, as is typical in the natural hazards literature, we use "hazard" for a natural event and "risk" for its effects on people and property.

The ways people address natural hazards can be grouped into four strategies:

- *Accept*: decide that the risk is not large enough to justify the economic, political, or other costs of action to reduce its potential effects.
- *Transfer*: use insurance or another method to pass the risk to someone else.
- *Avoid*: minimize exposure to the risk.
- *Mitigate*: take other measures to reduce damage and losses.

[1] Nathan Fick, One Bullet Away: The Making of a Marine Officer, 2006.

Playing against Nature: Integrating Science and Economics to Mitigate Natural Hazards in an Uncertain World, First Edition. Seth Stein and Jerome Stein.
© 2014 John Wiley & Sons, Ltd. Published 2014 by John Wiley & Sons, Ltd.
Companion Website: www.wiley.com/go/stein/nature

For example, we can address river flooding by doing nothing, insuring structures on the floodplain, banning development on the floodplain, or building levees to stop flooding. Each of these has advantages and disadvantages, costs and benefits.

Because there is no standard nomenclature, we could call all of these "mitigation," but it is useful to try to distinguish them while recognizing that they somewhat overlap. All of the possible measures involve where we live, what we build, and what we do. These are often combined into the idea of community *resilience* – the ability to prepare and plan for, absorb, recover from, and more successfully adapt to adverse events.

This book's goal is to suggest some general ideas about how to chose strategies to address hazards. What to do for a specific hazard in a particular place is a complex question to which there is no full, right or unique answer. However, some useful generalizations can be made. In this chapter we summarize key aspects of these strategies, and in the next chapter discuss ways of choosing among them.

11.2 Accepting Risk

People's approach to many risks is to accept them. We pretty much have to, because we are constantly being warned of dangers: sharks, SARS, West Nile virus, terrorists, computer viruses, Lyme disease, killer bees, communists, fiscal cliffs, flesh-eating bacteria, fire ants, mad cow disease, the end of the Maya calendar, alien abductions, genetically modified food, anthrax, etc. If we tried to address more than a small fraction of these, we would have no time or resources to do anything else. Fortunately, very few turn out to be anywhere near as serious as claimed. As we discussed in section 6.4 and Best (2004) points out in *More Damned Lies and Statistics*, "Apocalyptic claims do not have a good track record."

People accept a risk when the costs of addressing it appear greater than those associated with addressing it. However, some hazards are worth devoting resources to. Hazard and risk assessments should help individuals and communities to identify the most significant risks and decide the amount of resources to devote to addressing them.

For example, in the US, electric power outages are common after summer thunderstorms, and long outages occur after major storms like Hurricane Sandy. These outages are expensive, inconvenient, and in some cases dangerous. Most result from falling trees knocking down power lines. In contrast, power outages are much rarer in much of Europe, such as Germany or the Netherlands, where most power lines are underground and thus less prone to damage. The difference arises because US utilities have decided that the costs of "undergrounding" are too high. They thus accept the problem, which is mostly a problem for their customers, not them. As a result, many US home-owners and business are installing standby generators. This raises the question of whether it would be better to address this risk at a community, rather than individual, level.

11.3   Transferring Risk

Insurance transfers risk from property owners to insurers – usually insurance companies or governments. The basic idea of insurance is spreading the risk. For example, car insurance premiums are low because many drivers pay a premium each year, but only a few have accidents and file claims that need to be paid.

To see the advantage of spreading the risk, consider that the total annual worldwide losses from natural disasters in 2012 were about $170 billion (Figure 1.1), or about $25 for each person on earth. Although the risks are not that widely spread, insurance does spread risks.

Beyond helping individuals who suffer losses, insurance is important for communities because after a disaster insurance compensates property owners for damage and thus provides funds for rebuilding. As a result developed nations, where most of the losses are covered by insurance, recover much faster than developing nations where losses are uninsured.

Despite its value, insurance has limitations as a hazard mitigation method. First, it reduces risks to property, but not to lives. Second, insurance can be expensive. In the US, earthquake insurance policies typically only pay for damage that is more than 15% of a house's value, a level of damage which rarely happens. Given this deductible and the high cost of the insurance, 88% of homeowners in California – including many seismologists – do not have it, despite a state program to encourage it. In fact, the fraction of homeowners with earthquake insurance has been decreasing over time.

A major issue is that disasters could overwhelm insurance companies' ability to pay claims. Only a small number of homes in an area will burn down in any one year, and that number is about the same every year. In con-trast, a natural disaster can cause a huge number of claims at once in one area. Thus insurance companies have to build up huge financial reserves to prepare for rare events whose impact is hard to predict. This situation causes problems in hurricane-prone areas. Premiums for homeowners' insurance doubled in Florida between 2002 and 2007, tripling in some cases after the 2004–2005 hurricane season. Policyholders have had trouble collecting after hurricanes, and some insurance companies are refusing to sell more policies in areas such as Florida.

One proposed solution is government insurance for all hazards. New Zealand's Earthquake Commission insures property owners against earthquakes, landslides, volcanic eruptions, tsunamis, storms, flood, or resulting fires. An advantage of such programs is they cover people who could not afford insurance and are thus most impacted by a disaster. However, the costs are huge – the US federal flood insurance program was $18 billion in debt even before Hurricane Sandy.

Moreover, public insurance subsidizes development in dangerous places. Owning a home on a seaside barrier island prone to flooding and wind damage is a risky investment, but if it is insured, the owner risks only a small fraction of the home's value. Often, people collect insurance, rebuild in the same place, and collect again when disaster strikes. Although a private insurer would refuse to keep writing policies for these sites or charge very high premiums, a government insurer that does not have to make money will continue writing policies. This is good for the homeowner, but bad for society. It transfers risk to all the taxpayers, so people living in relatively safer areas subsidize those in more hazardous places. A successful national program would have to be designed carefully to avoid this problem.

11.4 Avoiding Risk

Land use planning reduces risks by preventing development in dangerous areas. After parts of Hilo, Hawaii, were seriously damaged by tsunamis in 1946 and 1960, the twice-damaged areas were redeveloped as parks. Development in dangerous areas near volcanoes and flood-prone areas along coasts and rivers can also be controlled. After a series of damaging floods, the Netherlands has adopted a strategy called "Room for the river" in which farmland is allowed to flood to absorb water that might otherwise flood populated areas. The program allows the Rhine River to carry much more water than previously without causing major problems.

Whether to prevent development is often a major political issue, with property owners fighting restrictions. The tradeoff is between the cost of taking valuable land out of use for many applications and the benefit of reduced damage if disaster strikes.

Evacuations are a short-term method of avoiding risks. Areas are sometimes evacuated in response to a threat of flooding or an imminent volcanic eruption. In many cases lives have been saved, but there are tradeoffs involved. Authorities face the challenge of deciding when to order evacuations, because of the difficulty in predicting how a storm or volcano will act (Chapter 5). In addition to the direct costs involved in an evacuation and the indirect costs of economic disruption, deaths have resulted from accidents as people flee. For example, in September 2005, as Hurricane Rita approached Texas, dire warnings induced 2.5 million people to evacuate coastal areas, creating 100-mile-long traffic jams in 100°F heat. As a result, only six of 113 deaths in Texas related to the storm were due to wind or water. The other 107 resulted from traffic accidents during the evacuation. Moreover, as Hurricane Katrina illustrated, older or poorer people may have difficulty evacuating unless adequate plans have been made. Careful planning and preparation can make evacuations smoother and safer.

Evacuations are crucial in tsunami hazard mitigation (section 3.2). Warning systems use seismological and geodetic information about a large earthquake that may generate a tsunami, and buoy systems that can detect the tsunami itself, to identify areas for evacuations. The Tohoku and Sumatra earthquakes provided impetus to improve these systems.

A special problem arises for tornadoes. Warning systems sometimes give people in small communities enough time to drive a short distance away and avoid the tornado. However, in congested cities where people need much longer to escape, people fleeing have been killed while stuck in traffic jams. Given that buildings are safer places to be than cars, deciding what to do is difficult.

11.5   Mitigating Risk

For simplicity, we consider all other approaches to reducing the effects of natural hazards as types of mitigation. Although the specifics vary between hazards, the general tradeoffs in deciding what to do are similar. Typically, they involve the costs and benefits of making buildings safer. For example, buildings in tornado-prone areas can be built with storm shelters. To date such communities have decided that the costs are too high – about $4,000 per home – to require shelters, but attitudes are shifting after a series of destructive tornadoes.

To explore the issues involved, we consider earthquake-resistant construction. Most earthquake-related deaths result when buildings collapse, although people standing in an open field during a large earthquake would just be knocked down. Thus an adage says that "earthquakes don't kill people, buildings kill people."

As discussed in section 10.7, ground shaking decreases with distance from the earthquake. Shaking at a given location is described by the acceleration, often as a fraction of "g", the acceleration of gravity ($9.8\,\mathrm{m/s^2}$). Shaking is also described in terms of *intensity*, a descriptive measure (Table 11.1) using

Table 11.1 The Modified Mercalli Intensity Scale

Intensity	
I	*Shaking not felt, no damage*: Not felt except by a very few under especially favorable conditions.
II	*Shaking weak, no damage*: Felt only by a few persons at rest, especially on upper floors of buildings.
III	*Felt quite noticeably by persons indoors, especially on upper floors of buildings*: Many people do not recognize it as an earthquake. Standing motorcars may rock slightly. Vibrations similar to the passing of a truck. Duration estimated.
IV	*Shaking light, no damage*: Felt indoors by many, outdoors by few during the day. At night, some awakened. Dishes, windows, doors disturbed; walls make cracking sound. Sensation like heavy truck striking building. Standing motor cars rocked noticeably. (0.015–0.02 g)
V	*Shaking moderate, very light damage*: Felt by nearly everyone; many awakened. Some dishes, windows broken. Unstable objects overturned. Pendulum clocks may stop. (0.03–0.04 g)
VI	*Shaking strong, light damage*: Felt by all, many frightened. Some heavy furniture moved; a few instances of fallen plaster. Damage slight. (0.06–0.07 g)
VII	*Shaking very strong, moderate damage*: Damage negligible in buildings of good design and construction; slight to moderate in well-built ordinary structures; considerable damage in poorly built or badly designed structures; some chimneys broken. (0.10–0.15 g)
VIII	*Shaking severe, moderate to heavy damage*: Damage slight in specially designed structures; considerable damage in ordinary substantial buildings with partial collapse. Damage great in poorly built structures. Fall of chimneys, factory stacks, columns, monuments, walls. Heavy furniture overturned. (0.25–0.35 g)
IX	*Shaking violent, heavy damage*: Damage considerable in specially designed structures; well-designed frame structures thrown out of plumb. Damage great in substantial buildings, with partial collapse. Buildings shifted off foundations. (0.5–0.55 g)
X	*Shaking extreme, very heavy damage*: Some well-built wooden structures destroyed; most masonry and frame structures destroyed with foundations. Rails bent. (more than 0.6 g)
XI	Few, if any (masonry) structures remain standing. Bridges destroyed. Rails bent greatly.
XII	*Damage total*: Lines of sight and level are distorted. Objects thrown into the air.

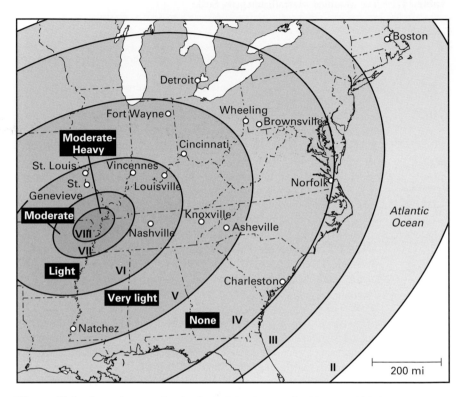

Figure 11.1   Intensity map for the first of the three major New Madrid shocks, on December 16, 1811. Boxes label different damage zones corresponding to intensity contours. Some of the sites from which reports exist are shown. (Stein, 2010. Reproduced with permission of Columbia University Press.)

Roman numerals ranging from I (generally unfelt) to XII (total destruction). Greater intensity values correspond to higher accelerations, although the relation is only approximate.

Intensity is inferred from human accounts, and so can be determined for earthquakes that occurred before the modern seismometer. For example, intensity data provide much of what is known about the New Madrid earthquakes of 1811 and 1812 (Figure 11.1). These large earthquakes are interesting in that they occurred in the relatively stable continental interior of the North American plate (Figure 9.6). Historical accounts show that although the earthquake was felt over a large area, damage was minor except near the tiny Mississippi river town of New Madrid. In St. Louis, a newspaper reported that "No lives have been lost, or has the houses sustained much injury. A few chimneys have been thrown down." Similar minor damage with "no real injury" occurred in Nashville, Louisville, Natchez, and Vincennes. No damage occurred in Fort Wayne, Wheeling, Asheville, Brownsville, Norfolk, or Detroit. The oft-repeated story of church bells ringing in Boston, however, turns out to be untrue. These data have been used to infer the magnitude (about 7.2 ± 0.3 in the study shown) of the historic earthquakes, which is important for assessing hazards (section 10.7). They also give insight into magnitude of, and shaking to expect from, future large earthquakes, provided – a large assumption – that these are similar to the past ones.

Today, people log on to websites like the USGS's "Did You Feel It?" after an earthquake and report the intensity they experienced, thereby allowing very detailed maps to be drawn up almost immediately after an earthquake. For example, a moderate (magnitude 5.2) earthquake in southern Illinois in 2008 generated almost 38,000 responses. Examples of such maps are shown in problem 11.10.

The damage resulting from a given ground motion depends on the types of buildings in the area. As shown in Fig 11.2, reinforced concrete (r-c) fares

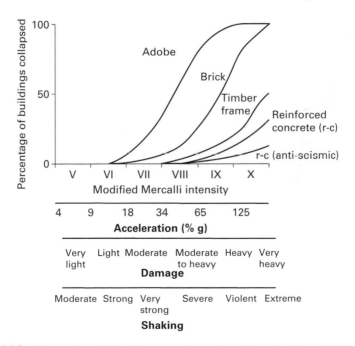

Figure 11.2   How vulnerable buildings are depends on the material used in their construction. (Stein and Wysession, 2003. Reproduced with permission of John Wiley & Sons.)

better during an earthquake than a timber frame, which does better than brick or masonry. Serious damage occurs for about 10% of brick buildings starting above about intensity VII–VIII (about 0.2–0.3 g), whereas reinforced concrete buildings have similar damage only around intensity VIII–IX (0.3–0.5 g). Buildings designed with seismic safety features do even better.

These values are just an average, for many reasons. How buildings fare depends on their design and quality of construction as well as what material they have been constructed from. Earthquake engineers have developed fragility functions that provide more detailed information and can be used to predict the fraction of buildings of various types that collapse at various shaking intensities. However, how a building responds to shaking depends on more than the intensity, which is just a simple way of describing the shaking. Both the duration of strong shaking and the period of the seismic waves also matter.

Society can thus reduce earthquake losses significantly by building earthquake-resistant buildings. The natural tendency among emergency planners and earthquake engineers is to favor more stringent building codes that require safer buildings. However, such design raises construction costs by about 3–10%, depending on a range of factors including the structure involved and the specifics of the design.

Retrofitting existing buildings to make them safer is possible, but very expensive. Figure 11.3 shows reconstruction of the Memphis, Tennessee,

Figure 11.3   The Memphis Veterans' Administration hospital during reconstruction and seismic retrofitting. (Courtesy of Joseph Tomasello.)

Veterans' Administration hospital, including seismic retrofitting, which cost $64 million. This involved removing nine floors of the 14-story tower, which made the cost about the same as constructing a new building. Retrofitting involves spending a large proportion of a building's cost in hopes of reducing the expected loss in a possible future earthquake. Whether this makes sense depends on the assumed probability of an earthquake during the building's life and the expected reduction in damage.

For example, a study of apartment buildings in earthquake-prone Istanbul, Turkey, found that the retrofit cost is about 40% of that of new construction. The reduction in the property loss if a large earthquake occurs was estimated to be 46% of the building's cost for sites shaken at intensity IX and 29% for sites shaken at intensity VIII. However, because most people will not pay more to live in a safer building, a retrofit would not raise building values significantly. The authors therefore concluded that these benefits did not justify the cost.

We will explore this topic in the next chapter. Deciding whether the costs are justified involves scaling the anticipated loss reduction if a large earthquake occurs by terms describing the probability that such an earthquake will happen during the building's remaining useful life. As we will see, the reduced property loss is unlikely to justify a retrofit unless we think the probability of an earthquake is high. However, retrofits can make sense in terms of saving lives if the earthquake hazard is high enough.

These measures make a community safer, but divert resources from other uses, some of which might save more lives at less cost or otherwise do more societal good. The challenge is to assess the seismic hazard and choose a level of earthquake-resistant construction that makes sense.

Ideally, building codes would be neither too weak, permitting unsafe construction and undue risks, nor too strong, imposing unneeded costs and encouraging their evasion. Deciding where to draw this line is a complex policy issue for which there is no unique answer. Making the appropriate decisions is even more difficult in developing nations, many of which face serious hazards but have even larger alternative demands for resources that could be used for seismic safety.

11.6   Combined Strategies

In many cases, a combination of strategies seems to be the best option. For example, in dealing with the danger of home fires, we accept some of the risk rather than incur the cost of fireproof houses. We transfer some of the risk by buying fire insurance, and mitigate some of the risk by paying for a fire department. We can mitigate the risk further, as discussed in the next chapter, by various measures including smoke detectors and sprinkler systems, to the extent that we decide that their benefits outweigh the costs.

Similar situations arise for other hazards. The storm surge from Hurricane Sandy did far less damage to those seaside communities in New Jersey and New York that had preserved or restored shoreline sand dunes. As a result, communities that had eliminated dunes to promote development are considering whether the cost and loss of beach access and views involved with restoring dunes are justified by their benefit in reducing damage. Coastal geologist Orrin Pilkey suggests a combined strategy: "If I was king, we would restore dunes, but we wouldn't rebuild destroyed homes close to the beach, and we'd move some buildings back anyhow. We would also put in regulations prohibiting intensification and development."

Questions

11.1.   Imagine that you are in charge of school construction in an earthquake-prone developing nation. How would you allocate your budget between building schools for towns without schools or making existing schools earthquake-resistant?

11.2.   Imagine that you are in charge of public works for a community that suffers frequent power outages after storms. The utility company is prepared to install underground power lines if the community pays the cost. How would you decide whether to do this?

11.3.   According to the *New York Times*, a Californian owning a home valued at $332,000 was offered earthquake insurance for $3,170 per year with a 10% deductible. Would you have bought this coverage? Why or why not?

11.4.   Would you favor the US having a mandatory national natural hazard insurance program analogous to New Zealand's? Why or why not? If yes, how should it be financed and operate?

11.5.   In September 2005, as Hurricane Rita approached Texas, dire warnings induced 2.5 million people to evacuate coastal areas, creating 100-mile-long traffic jams in 100°F heat. As a result, only six of 113 deaths in Texas related to the storm were due to wind or water. The other 107 resulted from traffic accidents during the evacuation. How should the risks of the evacuation process be factored into decisions as to whether to evacuate in cases like this?

11.6. If you were in charge of the Veterans' Administration, would you have carried out the seismic retrofit of the Memphis VA hospital (Figure 11.3) or used these funds otherwise? Explain your choice.

11.7. The owner of a large warehouse in Memphis, Tennessee, pointed out that the building is uninhabited, so any possible earthquake damage impacts only the owner. Hence instead of the government imposing a building code to protect against this damage, he proposed addressing the small risk involved through insurance, which would be much cheaper. Do you agree or disagree and why?

11.8. An example of a mitigation program not based on evidence is the fact that about 45 million Americans go to their doctors for annual physical examinations. However, many studies have found that these examinations have little or no benefit for people who do not have symptoms of disease. Estimate the cost of these examinations. Why do you think they remain so common? Should something be done about this, and – if so – what?

11.9. A major factor in tornado deaths in Oklahoma is that homes typically do not have basements, because of soil conditions. Communities have considered requiring storm shelters in new homes but have chosen not to. As a result, most homes do not have them. A builder who installs them in new homes said that he did not favor a requirement, because "the market ought to drive what people are putting in the houses, not the government." Do you agree or disagree, and why?

11.10. Shown below (and also as color plate Q11.10) are USGS maps of reported ground shaking, provided by David Wald, in the 1998 Northridge (M 6.7), and 2001 Nisqually (M 6.6), earthquakes. In these maps, stars show the epicenters. Northridge was the costliest earthquake in US history to date, with economic loss of about $40 billion. In contrast, loss in the Nisqually earthquake was about $2 billion. One death, a heart attack victim, was reported in the Seattle area, while 57 people died in the Northridge earthquake.
What are the shaking differences between the two similar sized earthquakes? What may have caused them?

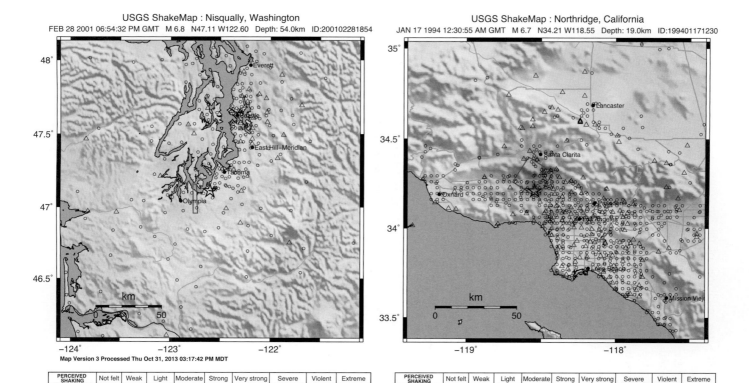

USGS ShakeMap : Nisqually, Washington

FEB 28 2001 06:54:32 PM GMT   M 6.8   N47.11 W122.60   Depth: 54.0km   ID:200102281854

Map Version 3 Processed Thu Oct 31, 2013 03:17:42 PM MDT

PERCEIVED SHAKING	Not felt	Weak	Light	Moderate	Strong	Very strong	Severe	Violent	Extreme
POTENTIAL DAMAGE	none	none	none	Very light	Light	Moderate	Mod./Heavy	Heavy	Very Heavy
PEAK ACC.(%g)	<0.05	0.3	2.8	6.2	12	22	40	75	>139
PEAK VEL.(cm/s)	<0.02	0.1	1.4	4.7	9.6	20	41	86	>178
INSTRUMENTAL INTENSITY	I	II–III	IV	V	VI	VII	VIII	IX	X+

Scale based upon Worden et al. (2011)

USGS ShakeMap : Northridge, California

JAN 17 1994 12:30:55 AM GMT   M 6.7   N34.21 W118.55   Depth: 19.0km   ID:199401171230

PERCEIVED SHAKING	Not felt	Weak	Light	Moderate	Strong	Very strong	Severe	Violent	Extreme
POTENTIAL DAMAGE	none	none	none	Very light	Light	Moderate	Mod./Heavy	Heavy	Very Heavy
PEAK ACC.(%g)	<0.05	0.3	2.8	6.2	12	22	40	75	>139
PEAK VEL.(cm/s)	<0.02	0.1	1.4	4.7	9.6	20	41	86	>178
INSTRUMENTAL INTENSITY	I	II–III	IV	V	VI	VII	VIII	IX	X+

Scale based upon Worden et al. (2011)

Further Reading and Sources

Morgan and Henrion (1990), Wilson and Crouch (2001), Vick (2002), and Hubbard (2009) discuss aspects of risk analysis and management. Carrns (2011) discusses earthquake insurance, including the example in problem 3. Information about the California and New Zealand insurance programs is at *http://www.earthquakeauthority.com* and *http://www.eqc.govt.nz*. For discussion of hurricane and flood insurance issues, see Michael-Kerjan and Kunreuther (2012), Kildow and Scorse (2012), and *http://www.nytimes.com/roomfordebate/2011/09/30/who-benefits-from-federal-flood-aid*.

Von Peter et al. (2012) compare the role of disaster insurance between developing and developed nations. The *Economist* (January 14, 2012; "Counting the cost of calamities") reviews "room for the river." Blumenthal (2005) and Carpenter et al. (2006) describe the Hurricane Rita evacuation. Hartocollis (2013) explains issues in evacuating hospitals and nursing homes for Hurricane Sandy. Schwartz (2013) discusses the tornado shelter issue.

The "Did You Feel It?" website is *http://earthquake.usgs.gov/earthquakes/dyfi/*. General explanations of earthquake-resistant construction are given in Gere and Shah (1984) and Levy and Salvadori (1992). More detailed discussions are in Coburn and Spence (2002) and FEMA (2006). Web resources include *http://www.celebratingeqsafety.com*. Jaiswal et al. (2011) present fragility functions for different building types.

Cost-benefit issues in earthquake hazard mitigation are discussed by Noll (1976), Stein et al. (2003), Stein (2004), Goda and Hong (2006), Searer et al. (2007), Crandell (2007), and Stein (2010). Retrofit issues are discussed by California Health Care Foundation (2007) for hospitals, Bernstein (2005) and Nagourney (2013) for older concrete buildings, and Erdik et al. (2003) for buildings in Turkey. Charlier (2003) describes the Memphis VA hospital reconstruction.

Navarro and Nuwer (2012) describe the role of sand dunes in protecting communities, and the Pilkey quotation is from Nuwer (2012).

References

Bernstein, S., How risky are older concrete buildings? *Los Angeles Times*, October 11, 2005.

Best, J., *More Damned Lies and Statistics: How Numbers Confuse Public Issues*, University of California, Berkeley, CA, 2004.

Blumenthal, R., Miles of traffic as Texans heed order to leave, *New York Times*, September 23, 2005.

California Health Care Foundation, Seismic Safety: will California's hospitals be ready for the next big quake, Issue Brief, January, 2007.

Carpenter, S., P. Campbell, B. Quiram, J. Frances, and J. Artzberger, Urban evacuations and rural America: lessons learned from Hurricane Rita, *Public Health Rep.*, Nov-Dec, *121*(6), 775–779, 2006.

Carrns, A., Is Earthquake Insurance Worth the Cost? *NYTimes.com*, September 7, 2011.

Charlier, T., Quake-proofing costly, difficult, *Memphis Commercial Appeal*, May 25, 2003.

Coburn, A. W., and R. J. S. Spence, *Earthquake Protection*, Wiley, New York, 2002.

Crandell, J., Policy development and uncertainty in earthquake risk in the New Madrid seismic zone, in *Continental Intraplate Earthquakes: Science, Hazard, and Policy Issues*, Special Paper 425, edited by S. Stein and S. Mazzotti, pp. 375–386, Geol. Soc. Amer., Boulder, CO, 2007

Erdik, M., E. Durukal, B. Siyahi, Y. Fahjan, K. Sesetyan, M. Demirciolu, and H. Akman, Instanbul case study, in *Earthquake Science and Seismic Risk Reduction*, NATO Science Series IV: Earth and Environmental Sciences, edited by F. Mulargia and R. J. Geller, pp. 262–283, Kluwer, Dordrecht, The Netherlands, 2003.

FEMA, *Designing for earthquakes*, Publication 454, 2006.

Gere, J. M., and H. C. Shah, *Terra Non Firma: Understanding and Preparing for Earthquakes*, W. H. Freeman, New York, 1984.

Goda, K., and H. P. Hong, Optimum seismic design considering risk attitude, societal tolerable risk level and life quality criterion, *J. Struct. Engineering*, *132*, 2027–2035, 2006.

Hartocollis, A., Health chief defends decision not to evacuate hospitals, *New York Times*, January 23, 2013.

Hough, S. E., J.G. Armbruster, L. Seeber, and J. F. Hough, On the Modified Mercalli Intensities and magnitudes of the 1811–1812 New Madrid, central United States, earthquakes, *Journal of Geophysical Research*, *105*, 23839–23864, 2000.

Hubbard, D. W., *The Failure of Risk Management*, Wiley, Hoboken, NJ, 2009.

Jaiswal, K., D. Wald, and D. D'Ayala, Developing empirical collapse fragility functions for global building types, *Earthq. Spectra*, *27*, 775–795, 2011.

Kildow, J., and J. Scorse, End federal flood insurance, *New York Times*, November 28, 2012.

Levy, M., and M. Salvadori, *Why Buildings Fall Down*, Norton, New York, 1992.

Michael-Kerjan, E., and H. Kunreuther, Paying for future catastrophes, *New York Times*, November 25, 2012.

Morgan, G. M., and M. Henrion, *Uncertainty: A Guide to Dealing with Uncertainty in Quantitative Risk and Policy Analysis*, Cambridge Univ. Press, Cambridge, UK, 1990.

Navarro, M., and R. Nuwer, Resisted for blocking the view, dunes prove they blunt storms, *New York Times*, December 3, 2012.

Nagourney, A., The troubles of building where faults collide, *New York Times*, November 30, 2013.

Noll, R., Defending against disaster, *Eng. Sci.*, May–June, 2–7, 1976.

Nuwer, R., Sand dunes alone will not save the day, *NYTimes.com Green Blog*, December 4, 2012.

Schwartz, J., Why no safe room to run to? Cost and Plains culture, *New York Times*, May 21, 2013.

Searer, G., S. A. Freeman, and T. F. Paret, Does it make sense from engineering and scientific perspectives to design for a 2475-year earthquake?, in *Continental Intraplate Earthquakes, Science, Hazard, and Policy Issues*, Special Paper 425. edited by S. Stein and S. Mazzotti, pp. 353–361, Geol. Soc. Amer., Boulder, CO, 2007.

Stein, S., No free lunch, *Seismol. Res. Lett.*, *75*, 555–556, 2004.

Stein, S., *Disaster Deferred: How New Science is Changing our View of Earthquake Hazards in the Midwest*, Columbia University Press, New York, 2010.

Stein, S., and M. Wysession, *Introduction to Seismology, Earthquakes, and Earth Structure*, Blackwell, Oxford, 2003.

Stein, S., A. Newman, and J. Tomasello, Should Memphis build for California's earthquakes? *Eos Trans. AGU*, *84*(177), 184–185, 2003.

Vick, S. G., *Degrees of Belief: Subjective Probability and Engineering Judgment*, Amer. Society of Civil Engineers, Reston, VA, 2002.

Von Peter, P., S. von Dahlen, and S. Sweta, Unmitigated Disasters? New Evidence on the Macroeconomic Cost of Natural Catastrophes, Working Paper 394, *Bank for International Settlements 2012*, Basel, Switzerland.

Wilson, R., and E. Crouch, *Risk-Benefit Analysis*, Harvard Univ. Press, Cambridge, MA, 2001.

12

Choosing Mitigation Policies

"It is our choices that show what we truly are, far more than our abilities."

J. K. Rowling, *Harry Potter and the Chamber of Secrets*[1]

12.1 Making Choices

As we have seen, society has a range of mitigation options for natural hazards. To choose between them, it needs to identify goals and then select strategies to achieve them. We operate under major constraints. First, we have only inadequate estimates about the occurrence and effects of future events. Second, we have limited resources to allocate between hazard mitigation and other needs. Third, we have a wide range of societal, political, and economic considerations. Given these factors, we have to decide somehow how much mitigation is appropriate – how much mitigation is enough.

To illustrate these complexities, consider motor vehicle accidents, which cause over 30,000 deaths per year in the US. Possible strategies to reduce the number of deaths include lowering speed limits, reducing drunk and otherwise distracted driving, improving roads, and making cars safer. Making cars safer is primarily an economic issue – how much we are willing to pay? However, reducing distracted driving – including that due to using cellular telephones and texting messages – involves both economic costs and our

---

[1]*Harry Potter and the Chamber of Secrets*: Copyright © J. K. Rowling, 1998.

Playing against Nature: Integrating Science and Economics to Mitigate Natural Hazards in an Uncertain World, First Edition. Seth Stein and Jerome Stein.
© 2014 John Wiley & Sons, Ltd. Published 2014 by John Wiley & Sons, Ltd.
Companion Website: www.wiley.com/go/stein/nature

willingness to change how we do things. Both go into deciding how much safety is enough.

We make similar decisions as individuals when we buy insurance for our homes, cars, or lives. The more we buy, the better we are covered, but the more it costs. That money cannot be used for other things. Thus instead of buying as much insurance as we can, we buy an amount that balances the costs and benefits based on our personal preferences.

Hazard mitigation amounts to a community doing the same thing, because the community is deciding how much "insurance" to buy against a possible future disaster. A thought experiment for scientists is to imagine being in a department in an earthquake-prone area that is planning a new building. The more of the budget they put into seismic safety, the better off they will be if an earthquake seriously shakes the building. However, the odds are good that the building will not be seriously shaken during its life. In this case, they are worse off because that money could have been used for office and lab space, equipment, etc. Economists call this the *opportunity cost* – the value of an alternative that was not pursued.

Deciding what to do involves cost-benefit analysis. This means trying to estimate the maximum shaking expected during the building's life, and the level of damage to accept if that happens. They have to consider a range of scenarios, each involving a different cost for seismic safety and a different benefit in damage reduction. They have to weigh these, accepting that estimates for the future have considerable uncertainties, and somehow decide on what they think is a reasonable balance between cost and benefit.

This process is what communities have to do in preparing for natural disasters. Although the final decision is never purely economic, good decision making involves carefully looking at costs and benefits of various options. Fundamentally, this involves two simple principles.

The first is "*there's no free lunch*." This means that resources used for one goal are not available for another. That is easy to see in the public sector, where there are direct tradeoffs. Funds spent strengthening schools against earthquakes cannot be used to hire teachers. Money hospitals spend on stronger buildings is not available to treat uninsured patients. Spending on stronger bridges can mean hiring fewer police officers and fire fighters.

A similar argument applies to saving lives. Safer buildings might over time save a few lives per year. On the other hand, the same money invested in other public health or safety measures (free clinics, flu shots, defibrillators, more police, highway upgrades, etc.) would save many more lives.

The second principle is "*there's no such thing as other people's money*." Often in planning hazard mitigation, the assumption is that someone else will pay. A government makes what is called an *unfunded mandate*, a rule that

someone else will pay to implement. Implicitly the idea is that rich developers, big businesses, or some other unspecified people will pay. This idea is
often even used to justify overestimating hazards. Unfortunately, it does not
work, because everyone ultimately pays. It is easy to see that safer public
buildings will cost taxpayers more. It is also true for the private sector,
because the costs for safer buildings affect the whole community. Some firms
will not build a new building, or will build it somewhere else. This reduces
economic activity and could cause job losses. In turn, state and local governments have less tax income for community needs: schools, parks, police, etc.
For those buildings that are built, the added costs get passed on directly to
the people who use the building – tenants, patients, clients, etc. – and then
indirectly to the whole community.

Who pays and how much gets tricky. For example, California has many
older concrete buildings, typically with weak first stories, that are not ductile
enough and so could collapse in an earthquake. Although it would be tempting
to require that these buildings should be strengthened, there are problems.
Many tenants would not or could not pay more for the extra safety, so property
owners could not pay the high cost and stay in business. Putting them out of
business would reduce the supply of inexpensive housing and office space.
The issue is that property owners would be told to pay for safety that does
not directly benefit them. They and their tenants would bear the costs, but the
benefit of safer buildings is to the whole community.

Economists call this issue, which often arises in natural hazard mitigation,
one of dealing with an *externality*. An externality is a cost or benefit that is
incurred by a party that was not involved as either a buyer or seller of the
goods or services causing the cost or benefit. In such cases, the community
has to decide how much safety it wants, how much it is willing to pay, and
how to pay for it.

To make things even more complicated, one person's cost can be another's
benefit. The cost of retrofitting – strengthening an existing building – is a cost
to the property owner but a benefit to the construction company. Similarly,
damage to a building in an earthquake is a cost to the owner or insurance
company, but a benefit to the company repairing it. These different interests
become important in the political decision making process.

As these examples show, deciding how much to spend on hazard mitigation
involves complicated choices. There is no unique or right answer. Sometimes
one hears claims that every dollar spent in hazard mitigation might save as
much as three dollars – or sometimes even ten dollars – in future losses.
Obviously, this cannot always be true. Some mitigation measures are very
cost effective, others are less so, and some cost much more than they could
save. To decide what to do, it is important to look carefully at the specific
situation and the costs and benefits of specific mitigation policies.

Surprisingly, most mitigation policies are chosen without such analysis. In general, communities have not looked at different options, and somehow end up choosing one even though they don't know how much they're paying or what they're getting for their money.

In fact, it turns out that hazard mitigation follows the general pattern of most government regulations. A joint study carried out by the Democratic-leaning Brookings Institution and the Republican-leaning American Enterprise Institute concluded that although the cost of federal environmental, health, and safety regulations was hard to measure, it was about as much as everything in the federal budget excluding defense and mandatory spending like Social Security. In their view:

> The benefits of these regulations are even less certain . . . Research suggests that more than half of the federal government's regulations would fail a strict benefit-cost test using the government's own numbers. Moreover, there is ample research suggesting that regulation could be significantly improved, so we could save more lives with fewer resources. One study found that a reallocation of mandated expenditures toward those regulations with the highest payoff to society could save as many as 60,000 more lives a year at no additional cost.

This situation is not unique to the US. Analogous regulations have been estimated to cost the 27 European Union countries an average of 3.7% of their gross domestic product. France, which is famous for such bureaucracy, is estimated to lose $10 billion per year and be "in danger of paralysis" as a result of rules that are designed to avoid all risk without consideration of their costs and benefits.

Obviously, this is not a good state of affairs, and we should do better.

12.2   House Fire Mitigation

To illustrate the issues in choosing mitigation strategies for rare natural disasters, we can consider protecting homes against fires, which are common enough that we have a sense of the issues involved.

We start by thinking about the hazard. In 2007, US fire departments responded to about 400,000 home fires, which killed about 3,000 people. Because there are about 300 million people in the US, the odds of being killed in a fire in any year are about 1 in 100,000. Because there are about 130 million residences, the chance of having a serious fire is about 1 in 300. A typical one of these fires causes about $18,000 in property losses.

These losses would be much higher if communities did not mitigate these risks by having fire departments, which cost taxpayers about $200 per person per year. In addition, banks require homeowners with mortgages to have insurance that covers fire damage, costing about $1000 per year. Insurance transfers the risk of property losses to the insurance company, but does not reduce the risk of being killed.

Homeowners can do more to mitigate the hazard. There are various options, which follow the usual pattern – the most effective systems cost more. The simplest option is to buy a fire extinguisher and smoke detectors. They are good, but only help when someone is at home. A better approach is to acquire a more expensive monitored alarm system, which operates even if no one is at home. The next level would be to install an automatic sprinkler system, which – compared to taking no action – would essentially eliminate the risk of death, and reduce property losses by about a third.

Studies have estimated and compared the costs and benefits of these options. We can use their results to make a table as a function of the mitigation level n.

Option 0 is to carry out no additional mitigation beyond that provided by the fire department and insurance. This option costs nothing, so its cost $C(0) = 0$. To estimate the expected property losses, we could say

$$L_p(0) = \text{probability of fire} \times \text{average uninsured loss in a fire}$$
$$= 1/300 \times \$18,000 \times 0.2 \times 1.1 = \$18, \quad (12.1)$$

assuming that 80% of the loss is covered by insurance, and indirect losses are about 10% of direct losses. This number is probably somewhat too low, because the $18,000 average loss includes some houses that had smoke detectors, alarms, or sprinklers.

Because saving lives is even more important, the analysis requires the uncomfortable placing of a value on potentially lost lives, as discussed in section 7.6. Assuming a life is valued at $5 million, following the practice of various government agencies, then if the average home has three people the expected annual human loss from a home fire is

Table 12.1   Cost-benefit appraisal of house fire mitigation options

n	Mitigation option	Cost ($)	Expected annual property loss	Expected annual human loss
0	None	0	$L_p(0)$	$L_h(0)$
1	Smoke detector	$C(1)$	$L_p(1)$	$L_h(1)$
2	Monitored alarm	$C(2)$	$L_p(2)$	$L_h(2)$
3	Sprinkler system	$C(3)$	$L_p(3)$	$L_h(3)$

$$L_h(0) = \text{probability of death in fire} \times \$5 \text{ million} \times 3$$
$$= 1/100,000 \times \$5,000,000 \times 3 = \$150. \tag{12.2}$$

This small amount reflects the fact that risk of death in a fire is low. Still, it is larger than the expected property loss, in part because most of the latter is insured.

To evaluate the benefits of different mitigation levels n, we consider the total (property plus human) annual expected reduced losses for each, relative to those expected for no mitigation ($n = 0$):

$$L_r(n) = [(L_p(n) + L_h(n)) - (L_p(0) + L_h(0))]. \tag{12.3}$$

We then multiply this annual expected reduced loss by the discount rate term to obtain the present value over T years, D_T, as discussed in section 7.5, to reflect the fact that the losses can occur in any year. Thus the net benefit of this level of mitigation is

$$D_T L_r(n), \tag{12.4}$$

which we compare to the cost of that level of mitigation, $C(n)$.

For example, assume a home sprinkler system reduces property losses by 1/3 and prevents any deaths, and that the insurance company gives a $50 discount because of the system being installed. Assuming a 5% interest rate over 30 years, $D_T = 15.4$, so the present value of the system's benefits is:

$$15.4 \times [(1/3 \times \$18) + \$150 + \$50] = \$3172. \tag{12.5}$$

Because sprinkler systems cost about $1500 to install in new houses, they make sense. As a result of such studies, some communities require them in new homes.

This example illustrates how we can carry out cost-benefit analyzes for hazards. In general, we compare the cost of a mitigation measure today to its expected benefit over time, which involves terms of the form

benefit = discount term × annual probability of event × reduced loss if

event happens = discount term × expected value of annual reduced loss.
$$\tag{12.6}$$

The equation shows the different factors involved. We multiply the assumed annual probability of an event by the anticipated loss reduction to find the expected value of the annual reduced loss. Thus more mitigation makes sense when we think the event is more likely or the loss reduction will be larger. Conversely, rarer events or smaller loss reduction justify less mitigation. Multiplying by the discount term includes the effect of time. For example, an event can occur at any time in a building's life. The longer the time considered, the more likely an event is, so a larger discount term indicates a larger benefit of mitigation.

Note that the reduced loss if an event occurs is multiplied by the product of the discount term and the annual probability of an event. For example, if the discount term is 15, and the probability is 1/100, their product is 0.15. Thus the expected benefit over time is only 15% of the reduced loss if an event happens.

Because mitigation makes sense when it costs less than its benefits, reducing its cost makes more mitigation cost-effective. Hence as mitigation technology improves and becomes cheaper, more mitigation makes sense. For example, sprinkler systems were installed first in large buildings, but are becoming cheap enough for homeowners and housebuilders to afford.

The fire safety example also illustrates some of the complexities involved in cost-benefit analysis. First, most of the numbers involved cannot be estimated well, either because of limited knowledge or because of the issues in valuing lives. Still, even rough estimates help in evaluating different options. For example, in the fire example they show that the primary gain is from reducing loss of life. If no lives are at risk, the benefit is mostly the reduced insurance premium, and is less than the system's cost. Hence sprinkler systems were first required in buildings where many people are exposed to the risk, such as apartment buildings or hotels. As sprinklers become cheaper, they can make sense in buildings where fewer people are at risk. Note that using a higher value for lives saved would make a sprinkler system even more cost effective.

Second, other factors are at work. For example, a monitored alarm system is primarily a burglar alarm system with sensors at doors and windows, so its cost cannot be directly compared with sprinklers.

Another issue involves the variability of the losses. Even if we have a good estimate of the average loss, there is a range of losses about this mean. Losses at the high end of the range are more of a problem for homeowners. As we will see in the next section, including this variability favors more mitigation.

There is also the familiar issue of how much is enough. As long as the benefits of more mitigation are greater than the costs, it makes economic sense to carry out that mitigation. However, even at some level of mitigation that makes sense compared to doing nothing, it may make still more sense to use the resources needed for another purpose, as we consider in the next section. Investment in mitigation is like any other – an investment that makes money is better than losing money, but money invested that produces a low yield could be better invested in ways that produce higher yield.

12.3 Losses from Hazards

Evaluating natural hazard mitigation strategies is conceptually similar to the fire example, but more complicated. Fires are relatively common, so the probability that one will happen and the resulting losses are relatively easy to estimate. However, because natural disasters are rare events, both their probabilities and the resulting losses are much harder to estimate. Still, conceptualizing the process gives insight into some of the factors involved. In many cases, even a simple and approximate comparison of costs and benefits can show whether a policy option is worth discarding or considering further.

For example, deciding whether to retrofit an existing building to make it more earthquake-resistant (Section 11.5) is conceptually similar to the sprinkler example, with the key difference that the probability of major earthquake damage is lower and harder to estimate. Even before doing a calculation for a specific area, the analogy suggests that the high costs involved are unlikely to be justified by the expected reduction in property loss over the building's remaining life, unless we think that the probability of an earthquake is high (equation 12.6). However, if the hazard is high enough they may be justified by the expected reduction in loss of life, especially in buildings where many lives are at risk.

We want to choose a level of mitigation that balances the reduction in the present value of expected losses against the cost of mitigation. As we have discussed, the benefits of saving lives are crucial but complicated to assess. Surprisingly, even defining the other losses, and thus the gain from avoiding them, is also quite complicated.

Consider a business that has had buildings or machinery destroyed in a disaster. We might first think that the loss is what it would cost to replace them. However, these assets could be generating a lot of income, or very little. In the first case, the loss would be much greater. Similarly, how much the firm collects from insurance or a government rebuilding fund need not correlate with the actual loss.

One of many ways to calculate the loss is to imagine that the firm's facilities are destroyed at time τ by a disaster. If it had been earning $y(\tau)$ dollars per year, the firm will lose $y(t)$ in annual income – profits, interest and dividends – over the period from $t = \tau$ to a future time $t = T$, when it fully recovers. The present value of this loss,

$$\sum_{t=\tau}^{T} y(t)/(1+i)^{t}, \qquad\qquad (12.7)$$

is the sum of money paid at time τ that would compensate the firm for the loss (section 7.5).

Because we do not know the future interest rate and the income lost in each of the T future years, we use an average annual loss from possible future disasters of $L(h,n)$. L depends on the scale of the natural event h, for example the height of a tsunami or the extent of ground shaking, and the level of mitigation n. The damage and thus loss of income increase with h and decrease with n.

As in section 7.5, the present value of this average loss is

$$L(h, n) \sum_{t=\tau}^{T} 1/(1+i)^{t} = L(h, n)/i \quad \text{for } T \text{ large.} \qquad (12.8)$$

To find the expected present value of the annual average loss of income from different events of scale h, we assume that the probability that an event of scale h will strike at time τ is $p(h)$. Therefore the expected present value of the annual average loss of income from an event at time τ is the sum

$$Q(n) = \sum_{h} p(h)L(h, n)/i. \qquad\qquad (12.9)$$

Because this estimate of the loss depends on the level of mitigation n, we can compare it to the cost of mitigation. Greater mitigation reduces the loss but costs more. How can we balance these to choose a mitigation strategy?

12.4 Optimal Natural Hazard Mitigation

Figure 12.1 shows a way to compare different options. For each, we define the cost of mitigation as $C(n)$, where n is a measure of mitigation. The scale of a natural event is parameterized by h, such as the height of a storm surge, the magnitude of an earthquake, or the level of the resulting ground shaking. The predicted economic loss, $L(h,n)$ increases with h and decreases with n. The probability of an event h is $p(h)$, so the expected present value of the expected loss is given by equation (12.9).

The optimum level of mitigation n^* minimizes the total cost, the sum of the present value of expected loss and mitigation cost:

$$K(n) = Q(n) + C(n) \qquad\qquad (12.10)$$

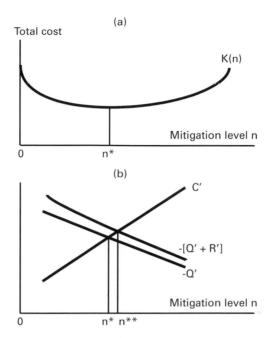

Figure 12.1   Comparing mitigation options. (a) The optimal mitigation level, n^*, minimizes the total cost, the sum of expected loss and mitigation cost. (b) n^* occurs when the reduced loss $-Q'(n)$ equals the incremental mitigation cost $C'(n)$. Including the effect of uncertainty and risk aversion, the optimal mitigation level n^{**} increases until the incremental cost equals the sum of the reduced loss and incremental decline in the risk term $R'(n)$. (Stein and Stein, 2012b. Reproduced with permission of the Geological Society of America.)

The U-shaped $K(n)$ curve illustrates the tradeoff between mitigation and loss. For no mitigation, $n = 0$, the total cost $K(0)$ equals the expected loss $Q(0)$ (Figure 12.1a). Initial levels of mitigation reduce the expected loss by more than their cost, so $K(n)$ decreases to a minimum at the optimum. $K(n)$ is steepest for $n = 0$ and flattens as it approaches the optimum, showing the decreasing marginal return on investment in mitigation. Relative to the optimum, less mitigation decreases construction costs but increases the expected damage and thus total cost, so it makes sense to invest more in mitigation. Conversely, more mitigation than the optimum gives less expected damage but at higher total cost, so the additional resources required would do more good if invested elsewhere.

The optimum can be viewed in terms of the derivatives of the functions (Figure 12.1b). Because increasingly high levels of mitigation are more costly, the marginal cost $C'(n)$ increases with n. Conversely, $-Q'(n)$, the reduced loss from additional mitigation, decreases. The lines intersect at the optimum, n^*, where $C'(n^*) = -Q'(n^*)$.

Naturally, there are major uncertainties in estimating both the probability of an event and the loss that would occur. The fact that the 2011 Tohoku tsunami was much greater than expected based on the Japanese earthquake hazard map showed that models that predict future occurrences of these events have large uncertainties. The failure of the seawalls to withstand the tsunami showed that the losses could be much greater than expected.

This analysis can include these uncertainties in several ways. The first is to assume that we have a good estimate of $K(n)$, the mean value of the expected loss, but to realize that there is a variance about this mean due to its uncertainty. As a result, the actual loss may be much greater than its mean. This is a problem, because we are risk averse in hazard mitigation. As discussed in section 7.4, risk aversion can be visualized using a game in which the probability of winning or losing a sum is the same, but we place greater weight on avoiding losing than on winning. Risk aversion is the ratio of the gain to the loss necessary to induce the player to bet, which is greater than one.

The combined effects of uncertainty and risk aversion can be included by adding a risk term $R(n)$ to the loss term $Q(n)$. Then $R(n) = \alpha\ \sigma^2(n)$ is the product of the risk aversion factor α and the variance of the estimated loss as a function of n. In this case, the optimum level of mitigation n^{**} minimizes the total cost, the sum of the expected loss and mitigation cost:

$$K(n) = Q(n) + C(n) + R(n). \qquad (12.11)$$

Using the derivatives shows that the mitigation level should be increased as long as $-[Q'(n) + R'(n)]$ exceeds the incremental cost of mitigation $C'(n)$, so the optimum mitigation level increases from n^* to n^{**}. $R'(n)$ is negative because the variance of the loss decreases with more mitigation – in the limit of enough mitigation there would be no loss.

Risk aversion differs from the other terms involved in that it is subjective. Different people in the same situation, facing the same risks, will buy different amounts of insurance. Similarly, different societies will invest different amounts to address a similar hazard, and the sum society will invest to mitigate different hazards often has little relation to the actual risk.

This makes risk aversion different from the other terms, which we can estimate. We can calculate the cost of mitigation $C(n)$ reasonably well. As discussed earlier, we estimate the probabilities $p(h)$ of hazards, although these estimates often have large uncertainties due to the variability of the earth's behavior. We can also estimate the mean and variance of the resulting loss $L(h,n)$ as a function of the hazard and mitigation level, although not very well. To pick just one problem, loss of life depends on the time of day, because people are in different places at different times.

The optimal mitigation level varies in space and time, depending on the expected loss $Q(n)$ and mitigation cost $C(n)$. This can be visualized by considering $K(0) = Q(0)$, the left intercept of the U-curve, which gives the expected loss in disasters without any mitigation. For areas where the expected loss is small, the left side of the curve shifts downward, so n^* shifts leftward, implying that less mitigation is optimum. Where expected losses are greater, such as in urban areas or for critical facilities like nuclear power plants, the left side of the curve shifts upward, so n^* shifts rightward, justifying higher mitigation.

The optimal mitigation level also increases if the cost of mitigation $C(n)$ declines, as often occurs as a result of improved technology or construction practices. The derivative $C'(n)$ shifts downward relative to $-Q'(n)$, so n^* shifts rightward, justifying higher mitigation. This has occurred in auto safety, where as systems such as antilock brakes become cheaper, they become cost-effective. The same process has been occurring for home fire sprinklers. One of the advantages of cost-benefit analysis is that it helps to identify mitigation measures that would be desirable if their costs can be reduced, which often catalyzes the necessary reduction.

An important variable to consider is the effects of economic growth. Consider two areas that are similar at $t = 0$. However, we anticipate that the economy of one area will grow at a rate $g > 0$, whereas that of the other will not. If the expected present value of the loss in the case of no growth is $Q(n) = \sum_h p(h)L(h, n)D_T$ (equation 12.9), then the present value of the anticipated loss from the hazard in the growing area is

$$Q(n) = \sum_h p(h)L(h, n)\sum_{t=0}^{T}(1+g)^t /(1+i)^t, \qquad (12.12)$$

where the summation is from now, $t = 0$, to a future time T. $K(0) = Q(0)$ is higher because of the multiplier $\Sigma(1 + g)^t/(1 + i)^t$, making the cost of no mitigation higher and favoring more mitigation there. Equivalently, the curve $-Q'(n)$ giving the incremental benefit from mitigation increases, raising the optimal level of mitigation.

Because different regions compete for limited resources in the budget, each will assert that its potential growth is greater and hence that it merits more resources for mitigation. Although it is hard to estimate the probability density function for the hazard $p(h)$, it can be even harder to agree on the pdfs for growth in different regions. Thus the uncertainty of projected growth may be an additional problem in formulating policy.

12.5 Nonoptimal Natural Hazard Mitigation

Graphs such as that in Figure 12.1 are schematic ways to guide our thinking, rather than functions we can compute exactly. It is conceptually useful to think about optimum mitigation strategies. However, given the uncertainties involved in estimating the quantities involved, it would be unrealistic to think that we can actually find an optimum strategy. Still, even simple estimates can show which strategies make more sense than others. Thus although in real cases these approaches cannot give an *optimum* strategy, they can identify *sensible* strategies.

Moreover, because mitigation policy decisions are made politically, they are unlikely to be the optimal decisions that would be made on purely economic grounds, even if we could find the optimum. Policies that make economic sense may not be politically viable. As we discussed in Chapter 6, society is sometimes overly concerned about relatively minor hazards and downplays others that are more significant. Hence in some cases we spend more than makes sense, and in others we spend less.

Figure 12.2 illustrates the effects of overmitigation and undermitigation. Choosing a level of mitigation n_u below the optimum n^* causes an excess loss from disasters equal to the difference between the total costs $K(n_u) - K(n^*)$. Graphically, this is the height of the U-curve above the dashed line for optimal mitigation. Similarly, choosing a level of mitigation n_o above the optimum n^* causes an excess cost of mitigation equal to $K(n_o) - K(n^*)$.

These nonoptimal solutions are inefficient, in that the optimum would be a better use of resources. However, a range of nonoptimal solutions is still

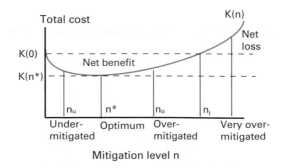

Figure 12.2 Illustration of the effects of overmitigation and undermitigation.

better than no mitigation. To see this, consider the dashed line $K(0) = Q(0)$ starting from the left intercept of the U-curve, which gives the total cost of no mitigation, the expected loss in disasters without any mitigation.

So long as we chose a mitigation level such that $K(n)$ is below this dashed line, the total cost is less than that expected if no mitigation is carried out, so there is a net benefit to society. The curve and line intersect on the right at a mitigation level n_l where

$$K(n_l) = Q(n_l) + C(n_l) = K(0). \qquad (12.13)$$

At this point, the benefit to society, the reduced loss compared to doing no mitigation, $Q(0) - Q(n_l)$, equals the mitigation cost $C(n_l)$. Higher levels of mitigation cost more than their benefit, and thus are thus a net loss to society, making them worse than no mitigation.

This analysis brings out some important points. First, in a situation where a hazard exists and no mitigation has been done yet, the initial modest levels of mitigation are almost certain to be better than doing nothing, because the U-curve typically decreases from $n = 0$. Second, a broad range of moderate mitigation levels are better than doing nothing. If they differ significantly from the optimal level, some of these resources could do more good if spent otherwise, but there is still a net benefit. Mitigation only does more harm than good in the extreme case of overmitigation where more is being spent to defend against disaster than the disaster would cost.

12.6   Mitigation Given Uncertainties

As we have seen, our ability to assess hazards and estimate losses in future disasters is limited. Thus we need to formulate policies while accepting the uncertainties involved. To see how this can be done, we consider a range of total cost curves between $K_1(n)$ and $K_2(n)$ (Figure 12.3). For simplicity, we view these as corresponding to high and low estimates of the hazard. They can also describe high and low estimates of the loss for a given event, or – most realistically – a combination of the uncertainties in hazard and loss estimates. These start at different values, representing the expected loss without mitigation. They converge for high levels of mitigation as the mitigation costs exceed the expected loss, because in the limit of enough mitigation there would be no loss.

In the limiting cases, the hazard is assumed to be described by one curve but is actually described by the other. As a result, the optimal mitigation level chosen as the minimum of the assumed curve gives rise to nonoptimal mitiga-

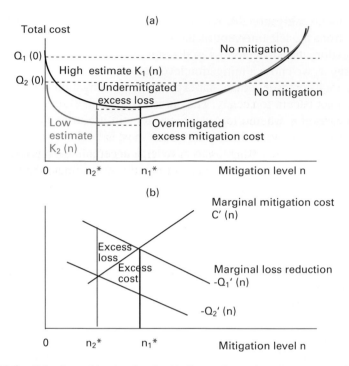

Figure 12.3   Selecting mitigation levels. (a) Comparison of total cost curves for two estimated hazard levels. For each, the optimal mitigation level, n^*, minimizes the total cost, the sum of expected loss and mitigation cost. (b) In terms of derivatives, n^* occurs when the reduced loss $-Q'(n)$ equals the incremental mitigation cost $C'(n)$. If the hazard is assumed to be described by one curve but is actually described by the other, the assumed optimal mitigation level causes nonoptimal mitigation, and thus excess expected loss or excess mitigation cost. (Stein and Stein, 2013b. Reproduced with permission of the Geological Society of America.) See also color plate 12.3.

tion, shown by the corresponding point on the other curve. Assuming low hazard when higher hazard is appropriate causes undermitigation and thus excess expected loss. This excess is shown by the height of the U-curve above the dashed line for optimum mitigation. In terms of the derivatives, it is the triangular area between the marginal loss reduction and marginal mitigation cost lines. Conversely, assuming high hazard when lower hazard is appropriate causes overmitigation and thus excess mitigation cost. However, so long as this point is below the dashed line for the correct curve, the total cost is less than expected from doing no mitigation.

Given the range of hazard estimates, we should somehow choose an estimate between them. The resulting curve will lie between the two curves, and thus probably have a minimum between n_1^* and n_2^*. Relative to the actual but unknown optimum, the resulting mitigation is thus nonoptimal, but perhaps not unduly so. In particular, so long as the total cost is below the actual loss for no mitigation, this nonoptimal mitigation is better than no mitigation.

For policy purposes, hazard assessments such as earthquake hazard maps and loss estimates are means, not ends in themselves. This analysis shows that inaccurate hazard and loss estimates are still useful as long as they are not too much of an overestimate. Given that most natural hazards assessments and estimates of the resulting losses have large uncertainties, it is encouraging to know that any estimate that does not greatly overestimate the hazard and loss leads to a mitigation strategy that is better than doing nothing.

12.7   Robust Policy Making

Once we recognize the large uncertainties in hazard and loss estimates, the challenge is how to choose an appropriate mitigation policy. Obviously, we cannot find an optimum policy given the uncertainties. The only way to even define an optimum is in hindsight, which is not useful. How, then, can we look for a sensible policy?

As scientists, our instinct is to try to reduce the uncertainties by getting more and better data and analyzing it in more sophisticated ways. Doing so can make a big difference. As we have discussed, using the historical and geologic record of large tsunamis in hazard assessment would have made a large difference for the Tohoku earthquake (section 2.2). In particular, the nuclear power plants would have been designed accordingly because such critical facilities should be designed to withstand rare low-probability events. Similarly, including geological and GPS data would have made a better earthquake hazard map for Haiti (section 1.2). Improved models that better predict ground motion can also reduce uncertainty in hazard estimates. Loss estimates can also be improved using better data for the building stock in an area and better models of the damage to expect in different cases.

However, we have seen that there are limits to how far uncertainties can be reduced. Because many reflect our limited knowledge of how natural processes work and the complexity of these processes, much of what we would like to know is poorly known or unknowable. Thus despite our best efforts as scientists, some of the uncertainties are unlikely to be reduced significantly on a short timescale.

This issue has not been generally recognized in dealing with natural hazards, where the tradition is to speak of "the" hazard as though it were known or knowable, making policy choice straightforward. However, there is growing recognition of this issue in other risk management applications.

In a recent review, Cox (2012) explains:

Some of the most troubling risk management challenges of our time are characterized by deep uncertainties. Well-validated, trustworthy risk models giving the probabilities of future consequences for alternative present decisions are not available; the relevance of past data for predicting future outcomes is in doubt; experts disagree about the probable consequences of alternative policies – or, worse, reach an unwarranted consensus that replaces acknowledgment of uncertainties and information gaps with groupthink – and policymakers (and probably various political constituencies) are divided about what actions to take to reduce risks . . . Passions may run high and convictions of being right run deep in the absence of enough objective information to support rational decision analysis . . .

The review cites examples of choosing policies to prepare the effects of climate change, manage risks from epidemics due to new pathogens, reduce new interdependencies in financial systems, and defend against cybercriminals. In each, one seeks policies to prepare for a very uncertain future.

To date, policies are usually chosen without rigorous analysis. However, a new approach under study is called *robust risk management* – accepting that there is a large range of possible scenarios and making policies accordingly. One accepts the uncertainty and develops policies that should give a reasonable outcome for a large range of the possible scenarios. In our application, a robust mitigation policy would give acceptable results for a wide range of the possible hazard and loss scenarios. The policy would be robust to the shallow and deep uncertainties. Rather than trying to find an optimal policy, this approach seeks a sensible policy.

This is quite different from choosing a particular scenario from the range, which cannot be done objectively, and making a policy based on that particular scenario. When a single scenario is chosen, it is usually done subjectively via a "consensus." As the Fukushima disaster (section 2.3), financial crisis (section 6.3), and many other examples show, the consensus approach often leads to errors due to "groupthink" (section 6.4), in which major uncertainties are downplayed or ignored.

Accepting the uncertainties involves a cultural change. For example, in the US the costs and benefits of proposed legislation are estimated by the Congressional Budget Office as though they were exactly known, despite the fact that these multiyear estimates have large uncertainties. In contrast, in Great Britain the analogous reports include uncertainty estimates. Even within the sciences, practices vary. As discussed in section 5.1, weather and climate forecasts recognize uncertainties, whereas "the" earthquake hazard is typically reported as if it were precisely known.

The alternative is to accept the uncertainty. For example, instead of defining "the" earthquake hazard by combining hazard scenarios using a logic tree that requires subjective weighting (Figure 10.15), the range of scenarios could be used to represent the uncertainty in our knowledge. Although individual researchers would likely favor specific scenarios, it would be more useful to say that we are reasonably confident that the earthquake hazard is in this range. For each scenario, various loss models can be computed to give a range of possible losses.

Once the uncertainties are accepted, robust policies can be developed. For example, economic models are widely used for policy decisions although their predictions are often poor. However, tests using over 1,000 different economic data series over time showed that a simple arithmetic average of forecasts made by different models usually outperformed any of the individual forecasts being averaged. Although makers of each model would have had arguments for that model, the average did better. Averaging reduced the error resulting from relying on any single model. It turns out that most models contributed – even the single best-fitting model is not perfect, and even poorly fitting models can contribute some information.

In this spirit, a variety of methods, called *model ensemble methods*, are being developed to make decisions based a set of model predictions, without assuming that any specific model correctly describes the real world. The general approach is to

i)   generate a range of scenarios consistent with the data and knowledge available;
ii) optimize or find a reasonable policy for each scenario; and
iii) combine the resulting policies in some way into the policy to follow.

A variety of methods have been suggested for how to combine the policies. One common theme is to examine what will happen for each policy if the assumed scenario is incorrect, and come up with a policy that yields the best – or at least a reasonable – outcome. As usual, there is no unique or right way to do this, but there are many interesting and sensible approaches.

Figure 12.3 lets us visualize how this process could be used for natural hazards. First, for each hazard scenario, we would derive a range of loss scenarios, which yield a range of many cost curves. For each of these, we can choose a range of risk aversion values. For each of these (hazard, loss, risk aversion) triplets of parameters, we derive an optimum mitigation level. Then, we examine the consequences – total cost – for all triplets assuming that the other cost curves were in fact correct. The cost difference from choosing a policy that proves incorrect is called the "regret." The results of this exercise can be used to derive a mitigation level or policy that should give a reasonable result in most cases. It will not be as good as if we had successfully predicted the hazard and loss, but will in most cases be better than if we had incorrectly predicted what would happen.

How to best do this is an active research area in risk analysis. Although such robust policy formulation has not yet been applied to natural hazard mitigation, we think there is a strong case for doing so. In our view, only good can come from recognizing the range of possible hazard and loss scenarios resulting from our limited knowledge, and considering how to formulate mitigation policy given the deep uncertainty. It amounts to accepting the need for humility in the face of the complexities of nature, and making policy accordingly.

Questions

12.1.   As discussed in the readings, debate about whether to retrofit buildings in Los Angeles has gone on for many years, without analysis of costs and benefits. Formulate an equation to assess whether retrofitting a building makes sense given the retrofit's cost, the building's remaining life, the probability of earthquake damage, the reduced damage due to the retrofit, and the lives saved. Explain your assumptions and the uncertainties involved. How would you apply this to the Los Angeles situation?

12.2.   Japanese economist Hajime Hori, emeritus professor of economics at Tohoku University, asked the authors after the Tohoku earthquake raised the issue of a similar tsunami to the south (section 1.1):

> What can we, and should we do, in face of uncertainty? Some say we should rather spend our resources on the present imminent problems instead of wasting them on things whose results are uncertain. Others say that we should prepare for future unknown disasters precisely because they are uncertain.

If you were a Japanese resident, how would you respond to Professor Hori's question?

12.3.   The destruction of the Fukushima nuclear plant prompted intense debate in Japan about whether to continue using nuclear power. There are clear economic benefits to using nuclear power rather than more expensive fossil fuels. Moreover, nuclear plants do not generate the carbon dioxide that causes global warming. However, there is obvious danger in operating nuclear plants in a nation with widespread earthquake and tsunami risks. Although non-nuclear plants would also be vulnerable, their loss would not pose the potential dangers associated with the destruction of a nuclear plant. If you were a member of a commission advising the Japanese government, how would you evaluate the issue?

12.4. After hospital buildings collapsed in the 1971 San Fernando earthquake, killing about 50 people, the state of California required that hospitals be strengthened. However, more than thirty years later, most of the state's hospitals do not meet the standard, and at least $50 billion would be needed to implement the upgrades necessary. There are additional problems because many hospitals are in poor financial shape, and 20% of Californians, about seven million people, have no health insurance and so rely on free care from hospitals. Such uninsured people are estimated to be 25–40% more likely to die than those who carry insurance. How would you decide on an allocation of resources between retrofitting for earthquake safety, and providing care for uninsured people?

12.5. Given the damage to New York City by the storm surge from Hurricane Sandy, possible options range from continuing to do little, through intermediate strategies such as providing doors to keep water out of vulnerable tunnels, to building barriers to keep the surge out of rivers. Progressively more extensive mitigation measures cost more, but are expected to produce increasing reduction of losses in future hurricanes. How would you develop a strategy to choose between the various proposed options? How would you include the anticipated but uncertain effects of global warming?

12.6. The ultimate natural disaster is the impact of an asteroid or short-period comet with the Earth, which could cause a global disaster or – if the object is large enough – a mass extinction of life like the one that ended the age of the dinosaurs. On the positive side, such impacts are the only major natural hazard that we can fully protect ourselves against, by detecting the object and deflecting it before it hits the earth. Review the status of present efforts in this direction, starting from a web search for "planetary defense." How would you evaluate these efforts? Should more or less be done, why, and how?

Further Reading and Sources

The report on the costs and benefits of federal regulations is Hahn and Litan (1998). Cody (2013) describes the analogous situation in Europe.

Our discussion of house fire mitigation draws on Butry et al. (2007). The optimal mitigation discussion follows Stein and Stein (2012, 2013).

How GPS data could have improved hazard assessment in Haiti is summarized in *http://www.usnews.com/science/articles/2010/01/15/scientists*

-warned-haiti-officials-of-quake-in-08 and the corresponding scientific paper is Manaker et al. (2008). Manski (2013) discusses uncertainty in government program costs. Morgan et al. (2009), Manski (2010, 2013), Cox (2012), and Hallegatte et al. (2012) discuss policy making given deep uncertainty. Navarro (2012) summarizes the discussion about protecting New York City from flooding. Retrofit issues for older concrete buildings in Los Angeles are discussed by Bernstein (2005) and Nagourney (2013).

References

Bernstein, S., How risky are older concrete buildings? *Los Angeles Times*, October 11, 2005.

Butry, D. T., M. H. Brown, and S. K. Fuller, *Benefit-Cost Analysis of Residential Fire Sprinkler Systems*, National Institute of Standards and Technology, Gaithersburg, MD, 2007.

Cody, E., France drowning in rules and regulations, critics say, *Washington Post*, April 16, 2013.

Cox, L. A., Jr., Confronting deep uncertainties in risk analysis, *Risk Anal.*, *32*, 1607–1629, 2012.

Hahn, R., and R. E. Litan, *An Analysis of the Second Government Draft Report on the Costs and Benefits of Federal Regulations*, AEI-Brookings Joint Center, Washington, DC, 1998.

Hallegatte, S., A. Shah, R. Lempert, C. Brown, and S. Gill, Decision making under deep uncertainty, Policy Research Working Paper 6193, World Bank, 2012.

Manaker, D. M., E. Calais, A. M. Freed, S. T. Ali, P. Przybylski, G. Mattioli, P. Jansma, C. Petit, and J. B. De Chabalie, Interseismic plate coupling and strain partitioning in the Northeastern Caribbean, *Geophys. J. Int.*, *174*, 889–903, 2008.

Manski, C. F., Vaccination with partial knowledge of external effectiveness, *Proc. Natl. Acad. Sci. U.S.A.*, *107*, 3953–3960, 2010.

Manski, C. F., *Public Policy in an Uncertain World*, Harvard University Press, Cambridge, MA, 2013.

Morgan, M. G., H. Dowlatabadi, M. Henrion, D. Keith, R. Lempert, S. McBride, M. Small, and T. Wilbanks, *Characterizing, communicating, and incorporating scientific uncertainty in climate decision making*, Report by the US. Climate Change Science Program and the Subcommittee on Global Change Research, 2009.

Nagourney, A., The troubles of building where faults collide, *New York Times*, November 30, 2013.

Navarro, M., New York is lagging as seas and risks rise, critics warn, *New York Times*, September 10, 2012.

Stein, J. L., and S. Stein, Rebuilding Tohoku: a joint geophysical and economic framework for hazard mitigation, *GSA Today*, *22*(9), 42–44, 2012.

Stein, S., and J. L. Stein, How good do natural hazard assessments need to be?, *GSA Today*, *23*(4/5), 60–61, 2013.

13

Doing Better

If you're gonna play the game, boy, ya gotta learn to play it right.
                                    Kenny Rogers, *The Gambler*[1]

13.1 Final Thoughts

This book started with experiences from around the turn of the century which show that, in its high-stakes game against nature, society is often not doing well. As we have seen, sometimes nature surprises us, when an earthquake, hurricane, or flood is bigger or has greater effects than expected from detailed natural hazard assessments. In other cases, nature outsmarts us, doing great damage despite expensive mitigation measures or causing us to divert limited resources to mitigate hazards that are overestimated.

We then explored what causes these problems and examined two approaches to doing better. First, scientists should try to do better at assessing future hazards, recognizing and understanding the deep uncertainties that are sometimes involved, and communicating these uncertainties to the public and planners formulating mitigation policies. Second, mitigation policies should be developed by considering both the uncertainties in the hazard and loss estimates and the costs and benefits of alternative strategies. Both approaches are challenging, but have the potential to significantly improve our ability to deal with natural hazards.

[1]Lyrics from "The Gambler," words and music by Don Schlitz, copyright © 1977 Cross Keys Publishing Co., Inc. All rights administered by Sony Music Publishing.

Playing against Nature: Integrating Science and Economics to Mitigate Natural Hazards in an Uncertain World, First Edition. Seth Stein and Jerome Stein.
© 2014 John Wiley & Sons, Ltd. Published 2014 by John Wiley & Sons, Ltd.
Companion Website: www.wiley.com/go/stein/nature

To address a hazard, we suggest starting from the basic questions in Chapter 7:

- What is the problem?
- What do we know and not know?
- What are we trying to accomplish?
- What strategies are available?
- What are the costs and benefits of each?
- What is a sensible strategy given various assumptions and uncertainty?

Doing such careful analysis of the costs and benefits of alternative strategies would be a major advance in natural hazard planning. The results could be used to help communities formulate sensible mitigation policies.

13.2   Community Decision Making

Because there are no unique or right strategies, we think the decision of which ones to use and in what combination should be made democratically by a community. This would differ from typical practice, in which these decisions are imposed on communities with little or no discussion of the benefits and costs of alternatives. As we have seen, this top-down approach driven by presumed "experts" often does not work well.

A more democratic approach has several advantages. It seems likely that, as Sarewitz et al. (2000) (section 5.4) argue,

> technical products of predictions are likely to be "better" – both more robust scientifically and more effectively integrated into the democratic process – when predictive research is subjected to the tough love of democratic discourse.

Moreover, sociocultural factors are important in deciding what to do. Ultimately, communities, not their national government, will bear most of the costs. As a result, they have to weigh the costs and benefits more carefully than does a national government that requires a policy but leaves it to communities to find most of the resources required. In many cases, the community may have more effective ways to use these resources that would do more good. It should be their decision how to balance costs and benefits and make the difficult tradeoffs between them.

An example of community involvement is the process used in Tennessee in 2003 to decide whether to strengthen building codes in the Memphis area. The US government was pressing for California-level earthquake standards, without having analyzed either the costs or benefits involved.

There was considerable discussion in the media. The *Memphis Commercial Appeal* explored the issue in news coverage and editorially. For example, it noted:

> Don't expect state and local governments to follow Washington's lead at the Memphis Veterans Medical Center, which is undergoing a $100 million retrofit to protect the building against the potential threat of an earthquake. Unlike the federal government, local governments are required to balance their budgets, and expenditures of that size would be hard to justify.

The relevant state agency then held public hearings in response to a request by the mayors of Memphis and surrounding Shelby County. Different stakeholders got the opportunity to give their views. Federal agencies and other supporters of the stronger code gave their traditional arguments. However, broader perspectives for the community that were not considered in developing the proposed code were also thoughtfully presented.

Mayor Willie Herenton of Memphis pointed out the need for careful thought:

> With the potential damage this code can cause to our tax base and economy we should have complete data and ample opportunity to make the decision. Our community has financial dilemmas that necessitate making wise choices. We cannot fund all of the things that are needed. Is it appropriate for us to spend the extra public funds for the enhanced level of seismic protection above and beyond our current codes? Would we be better off paying higher salaries for our teachers and building classrooms? This is the kind of public policy debate that should occur before new codes are warranted.

Bill Revel, the mayor of nearby Dyersburg, a major manufacturing community, expressed concerns about economic development. He explained that the increase in building cost would hurt the town's ability to attract industry because "they look at everything. They look at tax incentives. They look at the labor force. They look at the cost of construction. And this would be very detrimental to us."

Architect Kirk Bobo explained,

> I do believe that the added cost for seismic design is material. We do work all over the country, California included. And we have probably more than average knowledge of cost implications of seismic design. And I believe further that the impact of non-construction, direct construction-related economic impact, and the cost implications of that, is something that warrants a tremendous amount of analysis. I don't believe that has been done, and I think it's critical that it be done before making an informed decision.

Architect Lee Askew echoed these thoughts:

> There is a premium to constructing buildings with this advanced level of
> seismic design. It will affect structure, certainly. It will heavily affect some of
> the other systems including mechanical, electrical, plumbing, etc . . . As I assist
> these large national clients, I see how competitive it can be from one community
> to the next, one state to the next. If we are doing something to hobble ourselves,
> then we will certainly be disadvantaged.

Developer Nick Clark pointed out that the proposed costs for reducing earthquake losses were much greater than used to fight West Nile virus (which
caused four deaths in the past year in the county), despite the much lower
earthquake risk. More generally, he argued,

> We have needs in our community that are much greater in terms of the invest-
> ment in human life. We have an educational system that is broke, that needs to
> be fixed . . . This is about seismic issues in earthquakes. But there's an earth-
> quake that can happen in terms of the construction of our society if we cannot
> take care of our citizens and train them to be healthy and productive individuals.
> There needs to be a connection between how we handle these resources in terms
> of the investment in citizens of Shelby County . . .

We think the resulting decision, to modify the proposed code, was wise. More
importantly, regardless of the outcome, this discussion of costs, benefits, and
alternatives is a good model for other communities in formulating hazards policy.

13.3   Improved Organization

Improving the organizational structure for dealing with natural hazards would
also be worthwhile. This would help both in assessing hazards and developing
strategies to mitigate them, and in responding to the immediate effects of a
disaster and rebuilding after it. The latter two are beyond our scope here,
because they are primarily organizational rather than scientific issues, although
science can help, especially in disaster response via new methods that provide
rapid information about what has occurred and the expected damage. However,
they are also crucial and worth discussing.

Often after major disasters, it is clear that aspects of the hazard assessment
and mitigation process were not done well. As discussed in section 7.1, the
damage caused by Hurricane Katrina resulted from a hurricane protection
system that was inadequate because of poor planning and coordination among
the many federal, state, parish, and local agencies involved. Thus despite an
effective hazard assessment, disaster occurred.

The emergency response was also inadequate. The multiple agencies involved were led by the US's Federal Emergency Management Agency. FEMA, with 2,500 employees and a $6.5 billion annual budget, is part of the Department of Homeland Security. Prior to Hurricane Katrina, many Americans knew little of this organization. Its botched response to Katrina made FEMA famous. Director Michael Brown, who took the job after serving as commissioner of the Arabian Horse Association, knew less about what was going on than millions of TV viewers. When informed that thousands of people were trapped with little food and water in dangerous conditions in the city's convention center, Brown responded, "Thanks for update. Anything specific I need to do or tweak?" Although he did not go to the city or send help, he took care to be photographed with sleeves rolled up as a sign of activity. Firefighters from across the country who responded with their rescue gear, ready to go, were sent by FEMA to Atlanta for lectures on equal opportunity and sexual harassment.

Reconstruction was also handled poorly. FEMA housed people in trailers that made their inhabitants sick from poison fumes. Its bureaucracy became a national laughing stock, illustrated by the case of restocking New Orleans' aquarium. FEMA wanted the aquarium to buy fish from commercial vendors, and initially refused to pay when the aquarium staff caught the fish themselves, saving taxpayers more than half a million dollars. Simon Winchester, who wrote a history of the 1906 San Francisco earthquake, noted that the federal, state, and local governments responded better to that event, with no emergency management agencies, and using steam trains and horse drawn wagons, far better than they did to Katrina 100 years later with designated agencies, jet aircraft, helicopters, and television. Eight years after the flooding, many parts of the city have still not yet recovered.

Such problems are not uncommon after major disasters. Prior to 2010, Haiti's earthquake hazard was inadequately assessed (section 1.2). Because the country is very poor, little earthquake hazard mitigation had been done. After the earthquake, international, foreign national, and nongovernmental organizations offered an unprecedented level of help. However, much of the assistance was poorly planned, uncoordinated, and ineffective. Hence, as the *New York Times* explained three years later,

> despite billions of dollars spent – and billions more allocated for Haiti but unspent – rebuilding has barely begun and 357,785 Haitians still languish in 496 tent camps.
>
> More than half of the money has gone to relief aid, which saves lives and alleviates misery but carries high costs and leaves no permanent footprint –

tents shred; emergency food and water gets consumed; short-term jobs expire; transitional shelters, clinics and schools are not built to last.

Of the rest, only a portion went to earthquake reconstruction strictly defined. Instead, much of the so-called recovery aid was devoted to costly current programs, like highway building and H.I.V. prevention, and to new projects far outside the disaster zone, like an industrial park in the north and a teaching hospital in the central plateau.

Meanwhile, just a sliver of the total disbursement – $215 million – has been allocated to the most obvious and pressing need: safe, permanent housing. By comparison, an estimated minimum of $1.2 billion has been eaten up by short-term solutions – the tent camps, temporary shelters and cash grants that pay a year's rent.

'One area where the reconstruction money didn't go is into actual reconstruction,' said Jessica Faieta, senior country director of the United Nations Development Program in Haiti from 2010 through this fall.

Moreover, while at least $7.5 billion in official aid and private contributions have indeed been disbursed – as calculated by Mr. Clinton's United Nations office and by The Times – disbursed does not necessarily meant spent. Sometimes, it simply means the money has been shifted from one bank account to another as projects have gotten bogged down.

That is the case for nearly half the money for housing.

Similar problems remain in Japan, two years after the Tohoku earthquake. As the *Wall Street Journal* described in "Sluggish Tsunami Recovery Pits Rules Against Reason:"

Mounds of rubble and half-destroyed buildings blemish the landscape of tsunami-struck cities. Evacuated towns near the Fukushima Daiichi nuclear plant meltdown won't be inhabitable for decades.

Japan has a financial advantage over poorer nations stricken by recent natural disasters but its recovery has been bogged down by many of the same problems that delayed Haiti and Pakistan: an inflexible bureaucracy set up to prevent wasteful spending and a lack of leadership, local officials said.

The problem isn't money. . . . The problem, said Rikuzentakata Mayor Futoshi Toba, stems from government rules attached to the funding. Red tape entangled the city's first major project – public housing for the poor and elderly, and new police and fire stations. When builders sought to cut down trees on the property to begin construction, the city was asked by the Ministry of Agriculture, Forestry and Fisheries to wait six months to comply with forestry procedures.

Meanwhile, the Reconstruction Agency would approve using public funds for the new fire station only if it was a replica rebuilt in the same location. The bureaucratic thicket delayed groundbreaking on the project by more than a year.

The city also is spending time navigating laws and regulations with little practical application for the disaster-torn region. When the city wanted to build a supermarket, officials discovered they needed to apply for permission in a lengthy procedure because the land was designated for farm use.

"Japan is a country that really sticks to the rules, but there are emergency situations and this is one of those times," said Mr. Toba.

The reconstruction of L'Aquila, Italy, after the 2009 earthquake (section 5.4) also went poorly despite funding from both the Italian government and European Union. Four years after the earthquake, much of the historic town center was still uninhabitable. A European parliamentary committee found that the new housing built at high cost was "dangerous and unhealthy," in part because the contractors had "direct or indirect ties to organised crime."

These types of problems have led to suggestions that nations develop more effective methods to deal with natural hazards. One approach would be for governments to have an integrated natural hazard policy for earthquakes, floods, storms, wildfire, drought, etc. This would involve assessment of the different hazards and a strategy to mitigate and respond to them. The assessment would be based on up-to-date science in academia and government. Similarly, the mitigation and response plan would be developed jointly with state and local governments, universities, community groups, and the private sector, and consider the costs and benefits of alternative strategies. A consistent plan for different hazards would have many advantages over the situation in the US and many other countries, in which different hazards are differently and inconsistently handled by different organizations.

A good model might be that of national defense. As discussed in Chapter 7, defending a nation against enemies has many similarities to defending against natural hazards. For most nations, natural disasters cause much more damage and loss of life than military conflicts. However, most nations' structures for dealing with natural hazards are much less well developed than for defending against human enemies.

Defense planners have learned the hard way of the need for integrated organization and planning. As for natural hazards, a challenge is that many agencies including the various military and diplomatic services, intelligence agencies, health services, and economic and financial organizations are involved in various aspects of defense. Nations thus develop structures to look at a range of possible threats and plan for a response. In the US, this is done by the president's National Security Advisor and the National Security Council, the President's forum for considering national security and foreign policy matters with his senior national security advisors and cabinet officials. Since its formation in 1947, the council advises and assists the President on national security and foreign policies, and coordinates policies among government agencies. Many other nations have a similar organization.

A national natural hazard strategy could follow the precedent of the US Department of Defense, which conducts a major study every four years called the Quadrennial Defense Review. The QDR assesses likely requirements in the coming years and how US strategy, programs, and resources should be planned and allocated to meet them within the forecast budget plan. It recognizes that strategies should change as situations change, and considers budgetary constraints. Although no planning is perfect, the QDR process helped the military shift focus after the Cold War.

For such planning to be effective, it is important to consider a full range of options. Poor planning is likely to result when options are framed so as to force a specific choice. It has been said that foreign policy bureaucrats often present the President with three choices for dealing with a situation: unconditional surrender, nuclear war, or the option they favor.

National defense also shows that the best scientific advice comes from people at "arm's length." This was critical to the success of the US and British military research effort in World War II. Although their goals were military, the scientists and engineers were largely civilians and thus able to identify weaknesses, needs, and opportunities much better than had they been in the military chain of command. In the words of historian Stephan Budiansky, "they brought a scientific outlook and a fresh eye to problems that had often been dealt with until then only by tradition, prejudice, and gut feeling."

In our view, a similar process is now happening for natural hazards. New ideas, technologies, and approaches being developed by scientists, engineers, economists, risk analysts, and others around the world offer the prospect of greatly improving natural hazard assessment and mitigation. These improvements will help society fare better in its high-stakes game against nature.

Questions

13.1.   In 1848, pioneering economist John Stuart Mill noted, "What has so often excited wonder [is] the great rapidity with which countries recover from a state of devastation; the disappearance, in a short time, of all traces of the mischiefs done by earthquakes, floods, hurricanes, and the ravages of war." Do you think this is still true today? Why or why not? What has or has not changed and why?

13.2.   National systems for defending against human enemies are much more sophisticated than those for defending against natural disasters, although for most nations the latter do much more harm. Why do you think this is? Do you think this should this be changed? If so, how?

Further Reading and Sources

Much of this chapter is based on Stein (2010).

The *Memphis Commercial Appeal* editorial is from May 29, 2003, and quotations from the building code hearing on August 27, 2003 are from the hearing record.

A system that provides rapid information about the expected damage after an earthquake is described in *http://earthquake.usgs.gov/research/pager*.

Brown's "tweak" quotation appears in many news sources. The aquarium story is in the November 22, 2007 *New York Times*. The New Orleans, Haiti, Tohoku, and L'Aquila rebuilding stories are from Stein (2010), Sontag (2012), Wakabayashi (2013), and Segreti (2013). Kieffer (2013) explores improved natural hazard mitigation organization. Dyson (1981) and Budiansky (2013) discuss scientists' role in World War II.

References

Budiansky, S., *Blackett's War: The Men Who Defeated the Nazi U-Boats and Brought Science to the Art of Warfare*, Knoff, New York, 2013.

Dyson, F., *Disturbing The Universe*, Basic Books, New York, 1981.

Kieffer, S. W., *The Dynamics of Disaster*, W.W. Norton, New York, 2013.

Mill, J. S., *Principles of Political Economy*, 1848.

Sarewitz, D., R. Pielke, Jr., and R. Byerly, Jr., *Prediction: Science, Decision Making, and the Future of Nature*, Island Press, Washington DC, 2000.

Sontag, D., Rebuilding in Haiti lags after billions in post-quake aid, *New York*, December 23, 2012.

Segreti, G., European report slams misuse of quake funds, Financial Times, Nov 5 2013

Stein, S., *Disaster Deferred: How New Science is Changing Our View of Earthquake Hazards in the Midwest*, Columbia University Press, New York, 2010.

Wakabayashi, D., Sluggish tsunami recovery pits rules against reason, *Wall Street Journal*, March 8, 2013.

Index

Playing against Nature: Integrating Science and Economics to Mitigate Natural Hazards in an Uncertain World, First Edition. Seth Stein and Jerome Stein.
© 2014 John Wiley & Sons, Ltd. Published 2014 by John Wiley & Sons, Ltd.
Companion Website: www.wiley.com/go/stein/nature